THE ILLUSTRATED FLORA OF ILLINOIS

The Illustrated Flora of Illinois

ROBERT H. MOHLENBROCK, General Editor

ADVISORY BOARD:

Constantine J. Alexopoulos, *University of Texas*
Gerald W. Prescott, *University of Montana*
Aaron J. Sharp, *University of Tennessee*
Robert F. Thorne, *Rancho Santa Ana Botanical Garden*
Rolla M. Tryon, *The Gray Herbarium*

THE ILLUSTRATED FLORA OF ILLINOIS

FLOWERING PLANTS
willows to mustards

Robert H. Mohlenbrock

SOUTHERN ILLINOIS UNIVERSITY PRESS
Carbondale and Edwardsville

FEFFER & SIMONS, INC.
London and Amsterdam

Library of Congress Cataloging in Publication Data
Mohlenbrock, Robert H
   Flowering plants, willows to mustards.

   (The Illustrated flora of Illinois)
   Bibliography: p.
   Includes index.
   1. Botany—Illinois. 2. Dicotyledons. I. Title.
II. Series.
QK157.M623   582'.13'09773   79-10981
ISBN 0-8093-0922-X

This book is dedicated to
The Joyce Foundation
whose encouragement made this book possible

CONTENTS

ILLUSTRATIONS

FOREWORD

Since its conception in 1960, eight volumes of The Illustrated Flora of Illinois have been published. They include the volume on ferns and seven on flowering plants. Of these seven, five have been devoted to monocotyledonous plants and one thus far to dicotyledonous plants. This is the second volume treating dicots. Several additional volumes on dicots will follow, as well as works on algae, mosses, liverworts, lichens, fungi, and the genus *Carex*.

The concept of The Illustrated Flora is to present every group of plants known to occur in Illinois. A description is provided for each kind of plant, along with illustrations, showing the diagnostic features of each species. Distribution maps and ecological notes are included. Keys to aid in easy identification of the plants are presented.

An advisory board was created in 1964 to screen, criticize, and make suggestions for each volume of The Illustrated Flora of Illinois during its preparation. The board is composed of botanists eminent in their area of specialty—Dr. Gerald W. Prescott, University of Montana (algae); Dr. Constantine J. Alexopoulos, University of Texas (fungi); Dr. Aaron J. Sharp, University of Tennessee (bryophytes); Dr. Rolla M. Tryon, Jr., the Gray Herbarium (ferns); and Dr. Robert F. Thorne, Rancho Santa Ana Botanical Garden (flowering plants).

The author is editor of the series and will prepare many of the volumes. Specialists in various groups are preparing the volumes on plants of their special interest.

There is no definite sequence for publication of The Illustrated Flora of Illinois. Volumes will appear as they are completed.

The author is deeply grateful to the Joyce Foundation whose generous support made possible the production of this volume.

Robert H. Mohlenbrock

Southern Illinois University
February 14, 1979

THE ILLUSTRATED FLORA OF ILLINOIS

FLOWERING PLANTS
willows to mustards

WISCONSIN

IOWA

INDIANA

MISSOURI

KENTUCKY

County Map of Illinois

JO DAVIESS · STEPHENSON · WINNEBAGO · BOONE · McHENRY · LAKE
CARROLL · OGLE · DeKALB · KANE · COOK
WHITESIDE · LEE · DuPAGE
KENDALL · WILL
ROCK ISLAND · HENRY · BUREAU · La-SALLE · GRUNDY
MERCER · PUTNAM · KANKAKEE
STARK · MARSHALL · LIVINGSTON
WARREN · KNOX · PEORIA · WOODFORD · IROQUOIS
HENDER SON · McDONOUGH · FULTON · TAZEWELL · McLEAN · FORD
HANCOCK · MASON · LOGAN · DeWITT · CHAMPAIGN · VERMILION
SCHUYLER · MENARD · PIATT
ADAMS · BROWN · CASS · SANGAMON · MACON · DOUGLAS · EDGAR
PIKE · MORGAN · MOULTRIE · COLES
SCOTT · CHRISTIAN · SHELBY · CLARK
GREENE · CUMBERLAND
CALHOUN · JERSEY · MACOUPIN · MONTGOMERY
PAYETTE · EFFINGHAM · JASPER · CRAWFORD
MADISON · BOND · CLAY · RICHLAND · LAWRENCE
CLINTON · MARION · WAYNE · EDWARDS · WABASH
ST. CLAIR · WASHINGTON · JEFFERSON
MONROE · RANDOLPH · PERRY · HAMILTON · WHITE
FRANKLIN
JACKSON · WILLIAMSON · SALINE · GALL-ATIN
UNION · JOHNSON · POPE · HARDIN
ALEXAN DER · PULASKI · MASSAC

Introduction

All flowering plants may be divided into two great groups, the monocotyledons, or monocots, and the dicotyledons, or dicots. Monocots are plants which produce a single leaf, called a seed leaf, when the seeds first germinate. Dicots, on the other hand, are plants which give rise to a pair of seed leaves when germination occurs.

Although monocots are fewer in number than dicots, they include many important groups of plants. Grasses, sedges, lilies, orchids, irises, aroids, and pondweeds are some examples of monocots. The members of this group have been described in five previous volumes in The Illustrated Flora of Illinois.

Dicots, which are more numerous in Illinois, include well-known plants such as roses, peas, mustards, mints, nightshades, milkweeds, and asters. One volume treating dicots has been published previously in this series.

Since the time of Linnaeus, and even before, hundreds of attempts have been made to organize the orders and families of plants into a logical sequence, generally trying to place plants which have similar characteristics near each other. As more and more features are studied, aided recently by sophisticated instruments such as the scanning electron microscope, revised systems of classification are developed in an effort to depict a truer picture of relationships.

Most contemporary systems of classification have important features. No system can be said to be the best system since the phylogenists creating these systems must make their own value judgments concerning the importance of various characteristics.

I have chosen for The Illustrated Flora of Illinois a classification proposed by Robert Thorne in outline form in 1968. Thorne's concepts have been accepted generally in this work, although I have departed from his system in a few instances.

I am following Thorne in using the standard suffix -aceae for all families. Thus in this volume, the mustard family, traditionally known as the Cruciferae, becomes the Brassicaceae. Elsewhere in the Illinois flora, the Guttiferae becomes the Hypericaceae, the Leguminosae becomes the Fabaceae, the Umbelliferae becomes the Apiaceae, the Labiatae becomes the Lamiaceae, the Compositae becomes the Asteraceae, and the Gramineae becomes the Poaceae.

Since the Thorne classification is considerably different from the more traditional Engler system, the sequence for the dicots is presented next. Those names in boldface are described in this volume of The Illustrated Flora of Illinois.

Order Annonales
Family Magnoliaceae
Family Annonaceae
Family Calycanthaceae
Family Aristolochiaceae
Family Lauraceae
Family Saururaceae
Order Berberidales
Family Menispermaceae
Family Ranunculaceae
Family Berberidaceae
Family Papaveraceae
Order Nymphaeales
Family Nymphaeaceae
Family Ceratophyllaceae
Order Sarraceniales
Family Sarraceniaceae
Order Theales
Family Aquifoliaceae
Family Hypericaceae[1]
Family Elatinaceae
Family Ericaceae
Order Ebenales
Family Ebenaceae
Family Styracaceae
Family Sapotaceae
Order Primulales
Family Primulaceae
Order Cistales
Family Violaceae
Family Cistaceae
Family Passifloraceae
Family Cucurbitaceae
Family Loasaceae
Order Salicales
Family Salicaceae

Order Tamaricales
Family Tamaricaceae
Order Capparidales
Family Capparidaceae
Family Resedaceae
Family Brassicaceae
Order Malvales
Family Sterculiaceae
Family Tiliaceae
Family Malvaceae
Order Urticales
Family Ulmaceae
Family Moraceae
Family Urticaceae
Order Rhamnales
Family Rhamnaceae
Family Elaeagnaceae
Order Euphorbiales
Family Euphorbiaceae
Order Solanales
Family Solanaceae
Family Convolvulaceae
Family Polemoniaceae
Order Campanulales
Family Campanulaceae
Order Santalales
Family Celastraceae
Family Santalaceae
Family Loranthaceae
Order Oleales
Family Oleaceae
Order Geraniales
Family Linaceae
Family Zygophyllaceae
Family Oxalidaceae
Family Geraniaceae

[1] Called Clusiaceae by Thorne (1968).

Family Balsaminaceae
Family Limnanthaceae
Family Polygalaceae
Order Rutales
Family Rutaceae
Family Simaroubaceae
Family Anacardiaceae
Family Sapindaceae
Family Aceraceae
Family Hippocastanaceae
Family Juglandaceae
Order Myricales
Family Myricaceae
Order Chenopodiales
Family Phytolaccaceae.
Family Nyctaginaceae
Family Aizoaceae
Family Cactaceae
Family Portulacaceae
Family Chenopodiaceae
Family Amaranthaceae
Family Caryophyllaceae
Family Polygonaceae
Order Hamamelidales
Family Hamamelidaceae
Family Platanaceae
Order Fagales
Family Fagaceae
Family Betulaceae
Order Rosales
Family Rosaceae
Family Fabaceae
Family Crassulaceae
Family Saxifragaceae
Family Droseraceae
Family Staphyleaceae
Order Myrtales
Family Lythraceae
Family Melastomaceae

Family Onagraceae
Order Gentianales
Family Loganiaceae
Family Rubiaceae
Family Apocynaceae
Family Asclepiadaceae[2]
Family Gentianaceae
Family Menyanthaceae
Order Bignoniales
Family Bignoniaceae
Family Martyniaceae
Family Scrophulariaceae
Family Plantaginaceae
Family Orobanchaceae
Family Lentibulariaceae
Family Acanthaceae
Order Cornales
Family Vitaceae
Family Nyssaceae
Family Cornaceae
Family Haloragidaceae
Family Hippuridaceae
Family Araliaceae
Family Apiaceae[3]
Order Dipsacales
Family Caprifoliaceae
Family Adoxaceae
Family Valerianaceae
Family Dipsacaceae
Order Lamiales
Family Hydrophyllaceae
Family Boraginaceae
Family Verbenaceae
Family Phrymataceae[4]
Family Callitrichaceae
Family Lamiaceae
Order Asterales
Family Asteraceae

[2] Included in Apocynaceae by Thorne (1968).
[3] Included in Araliaceae by Thorne (1968).
[4] Included in Verbenaceae by Thorne (1968).

Three orders of vascular plants are included in this volume, encompassing five families. Because such a small number of families of dicots is found in this work, no overall key to the dicot families of Illinois is included. For keys to all families of vascular plants in Illinois, my companion volume, *Guide to the Vascular Flora of Illinois* (1975), is recommended.

The orders covered in this volume are the Salicales, Tamaricales, and Capparidales.

The Salicales and Tamaricales each are made up of a single family, the Salicaceae and Tamaricaceae, respectively. Three families comprise the Capparidales. These are the Capparidaceae, Resedaceae, and Brassicaceae.

The nomenclature for the species and lesser taxa used in this volume has been arrived at after lengthy study of recent floras and monographs. Synonyms, with complete author citation, which have applied to species in Illinois, are given under each species. A description, while not necessarily intended to be complete, covers the more important features of the species.

The common name, or names, is the one used locally in Illinois. The habitat designation is not always the habitat throughout the range of the species, but only for it in Illinois. The overall range for each species is given from the northeastern to the northwestern extremities, south to the southwestern limit, then eastward to the southeastern limit. The range has been compiled from various sources, including examination of herbarium material and some field studies. A general statement is given concerning the range of each species in Illinois. Dot maps showing county distribution for each taxon are provided. Each dot represents a voucher specimen deposited in some herbarium. There has been no attempt to locate each dot with reference to the actual locality within each county.

The distribution has been compiled from extensive field study as well as herbarium study. Herbaria from which specimens have been studied are located at Eastern Illinois University, the Field Museum of Natural History, the Gray Herbarium of Harvard University, the Illinois Natural History Survey, the Illinois State Museum, Knox College, the Missouri Botanical Garden, the Morton Arboretum, the New York Botanical Garden, Southern Illinois University at Carbondale, the United States National Herbarium, University of Illinois, and Western Illinois University. In addition, some private collections have been examined. The author is indebted to the curators and staffs of these herbaria for the courtesies extended.

Each species is illustrated, showing the habit as well as some of the distinguishing features in detail. Mr. Paul Nelson has prepared all of the illustrations.

Support for every aspect of this study has come from the generous contributions of the Joyce Foundation. To it and its directors, I am deeply grateful.

Many persons have contributed to the production of this work. Mr. Douglas M. Ladd has assisted me immeasurably in the field and in herbaria. Mr. George Fell and the Natural Land Institute are acknowledged for their cooperation. My wife Beverly has done all the clerical work involved in the project, from organizing my data to typing all drafts of the manuscript.

Descriptions and Illustrations

Order Salicales

The Salicales includes only the family Salicaceae. In the Thorne system of classification (1968), the Salicales is one of four orders comprising the superorder Cistiflorae. The other orders are the Cistales, Tamaricales, and Capparidales. Cronquist (1968) accords similar treatment, but recognizes the name Violales instead of Cistales (including within it the Tamaricales).

Even though the Salicales seems isolated from all other plant groups, Thorne and Cronquist both agree that it represents advanced derivatives of a cistalian or violalian ancestor, primarily on the basis of the compound, unilocular pistil with usually parietal placentation. Brown (1938) had earlier suggested a relationship among the Cistales, Salicales, and Capparidales on the basis of similar nectaries.

The earlier idea that the Salicales is very primitive has been refuted by most botanists today.

Only the following family comprises this order.

SALICACEAE – WILLOW FAMILY

Trees or shrubs, mostly dioecious; leaves simple, alternate (rarely nearly opposite), with stipules; flowers unisexual, arranged in catkins, without a perianth; staminate flowers each subtended by a bract and either a cupular disc or 1–2 glands, with 2 to several stamens, the filaments free or connate at the base, the anthers 2-celled, with vertical dehiscence; pistillate flowers each subtended by a bract and either a cupular disc or 1–2 glands, with one pistil, the ovary superior, unilocular, with 2–4 parietal placentae and numerous ovules, the style with 2–4 stigmas; fruit a capsule with 2–4 valves; seeds comose.

The Salicaceae is composed of two genera and about 350 species, distributed throughout most of the world. It differs from other catkin-bearing families by its dioecious nature, its bracteate flowers, and its comose seeds.

The absence of a perianth is interpreted by most botanists as a reduction from some ancestor having a perianth.

Many species of both *Salix* and *Populus* are grown as ornamentals.

KEY TO THE GENERA OF Salicaceae IN ILLINOIS

1. Leaves twice as long as broad or longer; bud scale 1; catkins not drooping
   _____ 1. *Salix*
1. Leaves never twice as long as broad; bud scales several; catkins drooping
   _____ 2. *Populus*

1. *Salix* L. – Willow

Dioecious trees or shrubs; buds with only one scale; leaves alternate, rarely subopposite; flowers unisexual, without a perianth; staminate flowers in erect or spreading aments, each flower subtended by one bract and 1–2 glands, with (1–) 2 (–10) stamens with free or basally connate filaments; pistillate flowers in erect or spreading aments, each flower subtended by a bract and 1–2 glands, with one pistil, the ovary superior, the 2 stigmas entire or 2-cleft; capsule 2-valved, with several comose seeds.

Salix differs from *Populus* by its elongated leaves and its buds with a single scale. In addition, all Illinois species of *Salix* have erect or spreading catkins, as opposed to the pendulous catkins of *Populus*.

There are about 300 different species of *Salix*, primarily in the north temperate and arctic regions of the world. In addition, a substantial number of hybrids has been reported.

Several species of *Salix* are common in cultivation. Included among these are the weeping willow (*S. babylonica* L.), the pussy willow (*S. discolor* Muhl.), the crooked willow (*S. matsudana* Koidz. var. *tortuosa* Hort.), the white willow (*S. alba* L.), the brittle willow (*S. fragilis* L.), the osier willow (*S. viminalis* L.), the purple osier (*S. purpurea* L.), and Thurlow's weeping willow (*S. elegantissima* Koch).

Several botanists have given considerable attention to the taxonomy of the genus. Foremost among those have been the treatments by Schneider (1921) and Raup (1943). In addition, Carlton R. Ball's studies on *Salix* have been highly respected in the United States.

Twenty-three species and two named hybrids are recognized in the Illinois flora. These fall into fifteen sections according to the treatment by Schneider (1921). Following is a summary of these sections and a list of the Illinois species in each.

§ Nigrae. Bracts falling away before capsule matures; stamens 3–6; flowers in whorls; pistillate flower with one gland; leaves green beneath. *Salix nigra*.

§ Bonplandianae. Bracts falling away before capsule matures; stamens 3–12; flowers in whorls; pistillate flower with one gland; leaves whitened beneath; stipules persistent; capsules granular. *Salix caroliniana*.

§ Triandrae. Bracts falling away before capsule matures; stamens 3–12; flowers in whorls; pistillate flower with one gland; leaves whitened beneath; stipules caducous; capsules smooth. *Salix amygdaloides*.

§ Pentandrae. Bracts falling away before capsule matures; stamens 3–12; flowers spirally arranged; pistillate flower with 2 glands. *Salix pentandra*, *S. lucida*, *S. serissima*.

§ Fragiles. Bracts falling away before capsule matures; stamens 2; catkins mostly solitary on scattered leafy shoots; leaves petiolate; pistillate flower with (1–) 2 glands. *Salix fragilis*, *S. babylonica*.

§ Albae. Bracts falling away before capsule matures; stamens 2; catkins mostly solitary on scattered leafy shoots; leaves petiolate; pistillate flower with 2 glands. *Salix alba*.

§ Longifoliae. Bracts falling away before capsule matures; stamens 2; catkins clustered on crowded branchlets; leaves sessile or nearly so. *Salix interior*.

§ Cordatae. Bracts persistent; stamens 2, the filaments free; staminate and pistillate flowers each with 1 gland; catkins flowering as the leaves unfold or after the leaves are half-grown, starting to flower from apex to base; ovary glabrous; stipules present; plants not stoloniferous. *Salix rigida*, *S. glaucophylloides*, *S. eriocephala*, *S. syrticola*.

§ Fulvae. Bracts persistent; stamens 2, the filaments free; staminate and pistillate flowers each with 1 gland; catkins flowering as the leaves unfold, starting to flower from base to apex; ovary silky; stipules present; plants not stoloniferous. *Salix bebbiana*.

§ Roseae. Bracts persistent; stamens 2, the filaments free; staminate and pistillate flowers each with 1 gland; catkins flowering as the leaves unfold; ovary glabrous; stipules absent; plants stoloniferous. *Salix pedicellaris*.

§ Discolores. Bracts persistent; stamens 2, the filaments free; staminate and pistillate flowers each with 1 gland; catkins flowering before the leaves begin to unfold; pedicel of capsule at least 3 times as long as the subtending gland; stipules not cordate, sometimes absent; stigmas elongate, slender. *Salix discolor*.

§ Griseae. Bracts persistent; stamens 2, the filaments free; staminate and pistillate flowers each with 1 gland; catkins flowering before the leaves begin to unfold; pedicel of capsule at least 3 times as long as the subtending gland; stipules not cordate, sometimes absent; stigmas short, thick. *Salix humilis*, *S. petiolaris*, *S. sericea*.

§ Capreae. Bracts persistent; stamens 2, the filaments free; staminate and pistillate flowers each with 1 gland; catkins flowering before the leaves begin to unfold; pedicel of capsule at least 3 times as long as the subtending gland; stipules cordate; plants not stoloniferous. *Salix caprea*.

§ Candidae. Bracts persistent; stamens 2, the filaments free; staminate and pistillate flowers each with 1 gland; catkins flowering before the leaves begin to unfold; pedicel not more than twice as long as the subtending gland; stipules not cordate; plants not stoloniferous. *Salix candida*.

§ Helix. Bracts persistent; stamens 2, the filaments united; staminate flowers each with 2 glands; catkins flowering before the leaves begin to unfold; leaves subopposite. *Salix purpurea*.

KEY TO THE TAXA OF Salix IN ILLINOIS

1. One or more glands present at upper end of petiole _____ 2
1. Glands absent at upper end of petiole _____ 4
   2. Leaves whitish beneath; capsule 7–10 mm long ____ 6. *S. serissima*
   2. Leaves green beneath; capsule less than 7 mm long _____ 3
3. Leaves with long-tapering, almost taillike, tip _____ 5. *S. lucida*
3. Leaves merely acute to short-acuminate at tip _____ 4. *S. pentandra*
   4. Leaves purplish, at least some of them opposite ___ 24. *S. purpurea*
   4. Leaves green or whitish, alternate _____ 5
5. Leaves glabrous on the lower surface _____ 6
5. Leaves pubescent, at least on the lower surface _____ 17
   6. Leaves remotely denticulate _____ 10. *S. interior*
   6. Leaves closely crenate or crenate-serrate or entire _____ 7
7. Leaves entire, revolute; some of the stems creeping _____
_____17. *S. pedicellaris*
7. Leaves crenate, serrate, or denticulate; all stems upright _____ 8
   8. Stipules absent or minute and falling away early on vegetative sprouts and young branchlets _____ 9
   8. Stipules persistent on vegetative sprouts and young branchlets _____
_____13
9. Leaves green beneath, coarsely undulate-serrate _____ 7. *S. fragilis*
9. Leaves whitish beneath, finely serrulate _____ 10
   10. Branchlets "weeping"; capsule 1.0–2.5 mm long _____
_____8. *S. babylonica*
   10. Branchlets not "weeping"; capsule 3–9 mm long _____ 11
11. Leaves tapering to a long-acuminate, almost taillike, tip _____
_____ 3. *S. amygdaloides*
11. Leaves acute to acuminate, but not taillike _____ 12

12. Flowers appearing before the leaves; teeth along margin of blade not extending all the way to the base _____ 20. *S. petiolaris*
12. Flowers appearing with the leaves; teeth along margin of blade extending all the way to the base _____ 9. *S. alba*
13. Leaves green on both sides _____ 14
13. Leaves pale on the lower surface _____ 15
14. Leaves lanceolate, rarely as much as 2 cm broad (except sometimes those on the sprouts), tapering to base; each staminate flower with two glands at the base; stamens 3 _____ 1. *S. nigra*
14. Leaves oblong-lanceolate, some of them at least 2 cm broad, rounded or even subcordate at base; each staminate flower with one gland at the base; stamens 2 _____ 11. *S. rigida*
15. Leaves irregularly crenate-serrate; flowers appearing before leaves expand; capsule puberulent _____ 18. *S. discolor*
15. Leaves finely serrulate; flowers appearing with the leaves; capsule glabrous or granular _____ 16
16. Leaves lanceolate, often falcate; each staminate flower with two glands at base; stamens 4–8; capsule 4–6 mm long, the pedicel at least half as long _____ 2. *S. caroliniana*
16. Leaves oblong to narrowly ovate, not falcate; each staminate flower with one gland at base; stamens 2; capsule 7–10 mm long, the pedicel much less than half as long _____
_____ 13. *S. glaucophylloides* var. *glaucophylla*
17. Young branchlets and new leaves covered with white wool _____
_____23. *S. candida*
17. Young branchlets and new leaves without white wool _____ 18
18. Leaves entire or undulate _____ 19
18. Leaves serrulate or crenate or denticulate _____ 21
19. Catkins appearing before the leaves; young branchlets densely tomentose or pilose _____ 19. *S. humilis*
19. Catkins appearing with the leaves; young branchlets sericeous, sparsely pilose, or glabrate _____ 20
20. Capsules up to 6 mm long; bracts brown, with a black tip _____
_____25. *S. × subsericea*
20. Capsules 6–10 mm long; bracts yellowish _____ 16. *S. bebbiana*
21. Leaves silvery-silky on lower surface _____ 22
21. Leaves pubescent beneath, but not with silvery-silky hairs _____ 25
22. Petiole up to 3 mm long; margin of blade remotely denticulate __
_____10. *S. interior*
22. Petiole 3 mm long or longer; margin of blade finely serrulate ____
_____23
23. Teeth along margin of blade not extending all the way to the base ____

_____20. S. *petiolaris*
23. Teeth along margin of blade extending all the way to the base ____ 24
   24. Young branchlets silky; flowers appearing with the leaves _____
   _____ 9. S. *alba*
   24. Young branchlets glabrous or glabrate; flowers appearing before
      the leaves _____ 21. S. *sericea*
25. Leaves narrowed or rounded at base, not subcordate; stipules on
   sprouts and young branchlets inconspicuous and falling away early
   _____ 26
25. Leaves subcordate at base (tapering in S. *eriocephala*); stipules on
   sprouts and young branchlets large and persistent _____ 29
   26. Young branchlets densely tomentose or pilose ____ 19. S. *humilis*
   26. Young branchlets sparsely pilose or glabrate _____ 27
27. Upper surface of leaves lustrous; flowers appearing before the leaves
   _____ 28
27. Upper surface of leaves dull; flowers appearing with the leaves _____
   _____16. S. *bebbiana*
   28. Capsule 7–12 mm long, minutely pubescent _____ 18. S. *discolor*
   28. Capsule 6–7 mm long, densely gray-hairy _____ 22. S. *caprea*
29. Branchlets permanently tomentulose _____ 30
29. Branchlets glabrous or glabrate _____ 31
   30. Leaves lustrous above, acute to short-acuminate _____
   _____15. S. *syrticola*
   30. Leaves dull above, tapering to a taillike tip _____
   _____ 14. S. *eriocephala*
31. Leaves glaucous beneath, tapering to a taillike tip _____
   _____12. S. × *myricoides*
31. Leaves green beneath, acute to short-acuminate _____ 11. S. *rigida*

1. **Salix nigra** Marsh. Arb. Am. 139. 1785. *Fig. 1.*
Salix falcata Pursh, Fl. Am. Sept. 2:614. 1814.
Salix nigra Marsh. var. *falcata* (Pursh) Torr. Fl. N. Y. 2:209.
1843.

Trees to 30 m tall, with a trunk diameter up to 1 m, the bark rough
and scaly, dark brown; twigs brownish or brownish-yellow, slender,
pubescent at first, at length glabrous, brittle near the base; leaves
linear to linear-lanceolate, acute to acuminate at the apex, rounded
to subcuneate at the base, straight or falcate, glandular-serrulate, to
10 cm long, to 1.5 cm broad, often pubescent when young, glabrous
or nearly so at maturity except for the puberulent midrib beneath,
green on both sides, the petiole to 8 mm long, puberulent, usually
with glands at the upper end; stipules (at least on the sprouts) glan-

1. Salix nigra (Black Willow). *a.* Flowering branch, X½. *b.* Leaf, X¾. *c.* Staminate flower, X10. *d.* Pistillate flower, X10. *e.* Fruiting raceme, X1. *f.* Seed, X15.

dular-serrate, auriculate, more or less persistent; aments borne before the leaves begin to expand; staminate aments erect to ascending, slender, to 10 cm long, the flowers whorled, each with 3–6 stamens, the filaments free, pilose near the base, the bract obovate, yellow, pubescent, persistent, the glands 2 or more; pistillate aments erect to ascending, slender, loosely flowered, to 6 cm long, the flowers whorled, each with a glabrous ovary, the bract oblong, yellow, pubescent, the gland one, the style minute or absent; capsules ovoid to ovoid-conical, 3–4 mm long, glabrous, on pedicels less than 1 mm long but over twice as long as the gland.

COMMON NAME: Black Willow.

HABITAT: Along streams, particularly in deep alluvial soil.

RANGE: New Brunswick to North Dakota, south to Texas and Florida.

ILLINOIS DISTRIBUTION: Common; in every county.

The black willow attains a greater size than any other species of Salix in the state. It frequently grows with several trunks.

Although the twigs are brittle near the base, they are tough and flexible above. The soft, light wood makes the black willow useful in the making of baskets, packing cases, and toys.

Salix nigra is perhaps confused most often with S. rigida or S. caroliniana. Salix rigida usually has some or all its leaves at least 2 cm wide, and lacks any glands on the petioles. Salix caroliniana looks very much like S. nigra, but is whitened on the lower leaf surface.

The flowers of the black willow open in April and May.

2. **Salix caroliniana** Michx. Fl. Bor. Am. 2:226. 1803. *Fig. 2.*
Salix longipes Shuttlw. ex Anders. Öfv. Svensk. Vetensk. Acad. Förh. 15:114. 1858.
Salix nigra Marsh. var. *wardi* Bebb ex Ward, Bull. U. S. Nat. Mus. 22:114. 1881.
Salix wardi (Bebb) Bebb, Gard. & For. 8:363. 1895.
Salix longipes Shuttlw. var. *wardi* (Bebb) Schneider, Bot. Gaz. 65:22. 1918.

Trees to 20 m tall, with a trunk diameter up to 0.5 m, the bark rough and scaly, usually gray or dark reddish-brown; twigs brownish, yellowish, reddish, or gray, pubescent, brittle near the base; leaves linear-lanceolate to oblong-lanceolate, acute to acuminate at the apex, rounded to subcuneate at the base, straight or falcate, glandular-serrulate, to 10 cm long, to 3 cm broad, more or less pubescent throughout, rarely glabrous, whitened on the lower surface, the petiole to 1 cm long, pubescent, usually with glands at the upper end; stipules (at least on the sprouts) glandular-serrulate, more or less auriculate, persistent; aments borne before the leaves begin to expand; staminate aments erect to ascending, slender, to 10 cm long, the flowers whorled, each with 4–8 stamens, the filaments free, pilose near the base, the bract obovate, yellow, villous, not persistent, the glands usually 2; pistillate aments erect to ascending, slender, rather densely flowered, to 10 cm long, the flowers whorled, each with a glabrous ovary, the bract oblong, yellow, villous, usually not persistent, the gland one, the style more or less absent; capsules conic to ovoid-conic, 4–6 mm long, granular, on pedicels up to 2 mm long, several times longer than the gland.

COMMON NAME: Ward's Willow.
HABITAT: Usually in gravel beds of streams.
RANGE: Maryland to Kansas, south to Texas and Florida; Cuba.
ILLINOIS DISTRIBUTION: Confined to the southern one-half of the state.
Although Bebb described this species in 1895 as *S. wardi*, the same species had been called *S. caroliniana* by Michaux in 1803. However, the common name Ward's willow is usually applied to this plant.

Engelmann discovered this species in Illinois in 1849, and it appears to have been rather common, at least in Madison and St. Clair counties. Today it is scattered but infrequently found in the southern one-half of the state.

Salix caroliniana differs primarily from *S. nigra* by its whitened lower leaf surface and from *S. amygdaloides* by its usually narrower leaves and its persistent stipules.

Although this species is thought to hybridize with *S. nigra*, *S. rigida*, and *S. sericea*, no such hybrids are definitely known from Illinois.

The flowers are borne in April and May.

3. **Salix amygdaloides** Anders. in Öfv. Svensk. Vetensk. Acad.
Förh. 15:114. 1858. *Fig. 3.*
Salix nigra Marsh. var. *amygdaloides* (Anders.) Anders. in DC.
Prodr. 16:201. 1868.

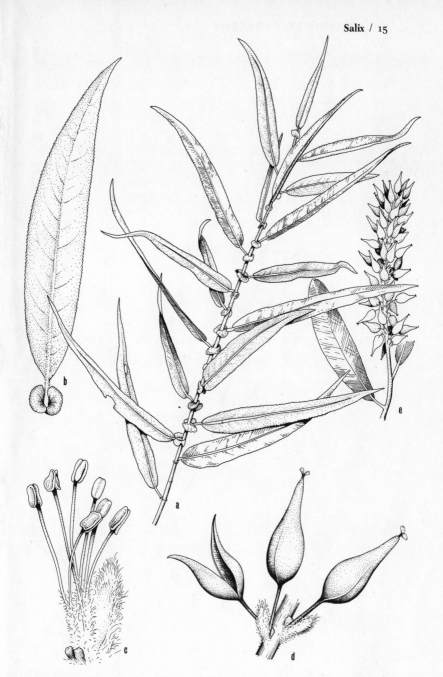

2. *Salix caroliniana* (Ward's Willow). *a*. Vegetative branch, X½. *b*. Leaf, X¾. *c*. Staminate flower, X7½. *d*. Pistillate flower and capsule, X7½. *e*. Fruiting raceme, X1.

3. *Salix amygdaloides* (Peach-leaved Willow).   *a*. Fruiting branch, X½.   *b*. Flowering branchlet, X¾.   *c*. Staminate flower, X10.   *d*. Pistillate flower, X10.   *e*. Seed, X20.

Trees to 20 m tall, with a trunk diameter up to 0.6 m, the bark rough and scaly, brown or reddish-brown; twigs yellowish or brown, slender, glabrous, tough and flexible; leaves lanceolate to ovate-lanceolate, attenuate at the apex, rounded or subcuneate at the base, straight, serrulate, to 15 cm long, to 2.5 cm broad, occasionally pubescent when young, glabrous on both surfaces at maturity, green above, glaucous below, the petiole to 2 cm long, glabrous, sometimes with glands at the upper end; stipules absent or minute, rarely persistent on sprouts; aments borne before the leaves begin to expand; staminate aments erect to ascending, slender, to 7 cm long, the flowers whorled, each with 3–5 stamens, the filaments free, pilose near the base, the bracts yellow, more or less pubescent, sparingly persistent, the glands 2; pistillate aments erect to ascending, slender, loosely flowered, to 8 cm long, the flowers whorled, each with a glabrous ovary, the bract oblong, yellow, more or less pubescent, the gland one, the style less than 1 mm long; capsules narrowly ovoid to ovoid-lanceolate, obscurely flattened, 3–4 mm long, glabrous, on pedicels 1.5–2.5 mm long, more than twice as long as the gland.

COMMON NAME: Peach-leaved Willow.

HABITAT: Along streams, low woods, borders of ponds.

RANGE: Vermont and Quebec to British Columbia, south to Washington, New Mexico, Texas, Missouri, and New York.

ILLINOIS DISTRIBUTION: Scattered throughout the state.

The peach-leaved willow differs from the somewhat similar S. caroliniana by its nonpersistent stipules and by its long-attenuate, broader leaves.

The lateral branches of S. amygdaloides tend to be more or less pendulous, and are exceedingly tough and flexible.

The flowers are borne from April to June.

Specimens reputed to be hybrids between Salix amygdaloides and S. nigra, called S. × glatfelteri Schneid., have been collected from Madison and St. Clair counties.

4. **Salix pentandra** L. Sp. Pl. 1016. 1753. *Fig. 4.*

Trees to 7 m tall, with a trunk diameter up to 30 cm (in Illinois); twigs brown to reddish-brown, slender, glabrous, shiny; leaves lanceolate to elliptic-oblong, acute to acuminate at the apex, rounded or subcuneate at the base, straight, glandular-serrulate, to 10 cm long, to 3 cm broad, glabrous, green above, green or a little pale be-

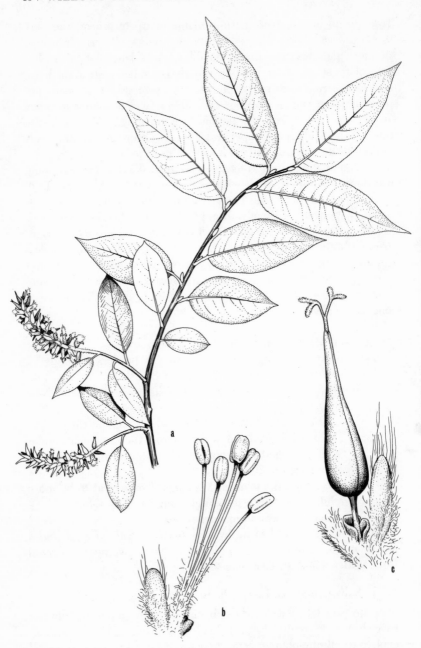

4. Salix pentandra (Bay-leaved Willow). *a.* Fruiting branch, X½. *b.* Staminate flower, X10. *c.* Pistillate flower, X10.

low, the petiole to 1 cm long, glabrous, with glands at the upper end; stipules usually minute, not persistent; aments borne as the leaves begin to expand; staminate aments erect to ascending, slender, to 6 cm long, the flowers spirally arranged, each with 5 stamens, the filaments free, pilose near the base, the bract obovate, yellow, more or less pubescent, the glands 2, often united; pistillate aments erect to ascending, slender, densely flowered, to 6 cm long, the flowers spirally arranged, each with a glabrous ovary, the bract oblong, yellow, pubescent, the glands 2, free, often cupular, the style about 1 mm long; capsules conical, 5–6 mm long, glabrous, on pedicels about twice as long as the gland.

COMMON NAME: Bay-leaved Willow.
HABITAT: Along a creek (in Illinois).
RANGE: Native of Europe; rarely escaped from cultivation in North America.
ILLINOIS DISTRIBUTION: Known only from Champaign, Lake, and Winnebago counties.
The bay-leaved willow is rarely found in North America as an escape from cultivation. At its Champaign County location, this willow grows along a creek near the Urbana railroad station. In Lake County, it has been found north of Libertyville.
The relatively short capsules distinguish this species from *S. serissima*, while the short fruiting pedicels distinguish it from *S. lucida*.
Salix pentandra flowers during May and June.

5. **Salix lucida** Muhl. Neue Schrift. Ges. Nat. Fr. Berlin 4:239. 1803. *Fig. 5*.
Salix lucida var. *intonsa* Fern. Rhodora 6:2. 1904.

Shrubs or trees to 7 m tall, with a trunk diameter up to 30 cm; twigs reddish-brown or yellowish, slender, glabrous or nearly so, shiny; leaves lanceolate to ovate-lanceolate, acute to acuminate to long-attenuate at the apex, rounded to subcuneate at the base, straight, glandular-serrulate, to 15 cm long, to 4 cm broad, sometimes pubescent when young, glabrous at maturity, green and shiny above, green or pale beneath, the petiole to 1.5 cm long, more or less glabrous, with glands at the upper end; stipules (at least of the sprouts) rounded, glandular, up to 6 mm long; aments borne as the leaves begin to expand; staminate aments erect to ascending, slender to thickish, to 4 cm long, the flowers spirally arranged, each with 3–6

5. *Salix lucida* (Shining Willow).   *a.* Vegetative branch, X½.   *b.* Staminate flower, X7½.   *c.* Pistillate flower, X7½.   *d.* Fruiting branchlet, X1.

stamens, the filaments free, pilose near the base, the bract oblong, yellow, pubescent, persistent, the glands 2; pistillate aments erect to ascending, rather slender, to 5 cm long, the flowers spirally arranged, each with a glabrous ovary, the bract oblong, yellow, pubescent, rather persistent, the glands 2, the style less than 1 mm long; capsules narrowly ovoid, 5–7 mm long, glabrous, on pedicels at least four times as long as the gland.

COMMON NAME: Shining Willow.

HABITAT: Bogs, moist disturbed areas.

RANGE: Labrador to Manitoba, south to Nebraska, northern Illinois, and Delaware.

ILLINOIS DISTRIBUTION: Confined to the northern one-fourth of the state.

The shining willow is most nearly related to S. *serissima* and the introduced S. *pentandra*. It differs from S. *serissima* by its larger stipules, long-attenuate leaves, and short capsules. It differs from S. *pentandra* by its larger stipules and long-attenuate leaves.

Tehon (1942) has reported var. *intonsa*, but this more pubescent variant seems to be unworthy of official taxonomic status.

In the bogs of northeastern Illinois, Swink (1974) records the shining willow growing in association with *Calopogon pulchellus*, *Drosera intermedia*, *Pogonia ophioglossoides*, *Rhus vernix* and other species.

Salix lucida flowers during May.

6. Salix serissima (Bailey) Fern. Rhodora 6:6. 1904. *Fig. 6.*
Salix lucida Muhl. var. *serissima* Bailey ex Arthur, Bailey, & Holway, Bull. Geol. & Nat. Hist. Surv. Minnesota 3:19. 1887.

Shrubs to 4 m tall; twigs yellowish to reddish-brown, slender, glabrous, shiny; leaves lanceolate to oblong-lanceolate, short-acuminate at the apex, rounded to subcuneate at the base, straight, glandular-serrulate, to 10 cm long, to 3.5 cm broad, glabrous, green above, whitened below, the petiole to 1 cm long, glabrous, with glands at the upper end; stipules minute or absent; aments borne after the leaves have begun to expand; staminate aments erect to ascending, to 3 cm long, the flowers spirally arranged, each with 4–7 stamens, the filaments free, pilose near the base, the bract oblong, yellow, pubescent, persistent, the glands 2, more or less united; pistillate aments erect to ascending, to 4.5 cm long, the flowers spirally arranged, each with a glabrous ovary, the bract oblong, yellow, pubescent, not persistent, the glands 2, the style up to 1 mm long; capsules conical-subulate, 7–10 mm long, glabrous, on pedicels at least twice as long as the gland.

COMMON NAME: Autumn Willow.

HABITAT: Bogs.

RANGE: Newfoundland to Alberta, south to Colorado, northeastern Illinois, and New Jersey.

ILLINOIS DISTRIBUTION: Known only from Lake and McHenry counties.

This extremely rare willow is closely related to S. *lucida*, but differs by its larger capsules, its short-acuminate leaves, and its minute stipules.

The common name autumn willow refers to the late dehiscence of the capsule, the seeds being liberated from mid-July to late August. The flowers are borne in early June.

7. **Salix fragilis** L. Sp. Pl. 1017. 1753. *Fig. 7.*

Trees to 25 m tall, with a trunk diameter up to 0.5 m, the bark thick, rough, gray; twigs yellow, brown, or gray, slender, glabrous or pubescent, shiny, very brittle at the base; leaves lanceolate, long-acuminate, cuneate at the base, straight, coarsely serrate, to 15 cm long, to 2.5 cm broad (sometimes much larger on sprouts), glabrous, green above, green or whitened below, the petiole to 1.5 cm long, glabrous, with glands at the upper end; stipules minute, caducous; aments borne as the leaves begin to expand; staminate aments erect to ascending, slender, to 6 cm long, the flowers spirally arranged, each with 2 (–4) stamens, the filaments free, pilose at the base, the bract oblong, yellow, sparsely pubescent, caducous, the glands 2; pistillate aments erect to ascending, slender, to 7 cm long, the flowers spirally arranged, each with a glabrous ovary, the bract oblong, yellow, sparsely pubescent, caducous, the glands 2, the style up to 1 mm long; capsules conical-subulate, 4–5 mm long, glabrous, on pedicels about twice as long as the uppermost gland.

COMMON NAMES: Crack Willow; Brittle Willow.

HABITAT: Along streams, near swamps.

RANGE: Native of Europe; escaped from cultivation from Newfoundland to Ontario and Minnesota, south to Missouri and Virginia.

ILLINOIS DISTRIBUTION: Scattered throughout the state.

The crack willow is one of six species of willow in the Illi-

6. *Salix serissima* (Autumn Willow). *a*. Fruiting branch, X½. *b*. Staminate flower, X5. *c*. Pistillate flower, X5. *d*. Capsule, X5.

7. *Salix fragilis* (Crack Willow). *a*. Flowering branch, X½. *b*. Leaf, X¾. *c*. Staminate flower, X10. *d*. Pistillate flower, X10. *e*. Immature fruiting branchlet, X1.

nois flora which is native to Europe and Asia, and is probably the most widespread of these as an introduction.

The branchlets are very brittle and break off readily. These broken branches easily take root in moist soil so that the species may spread rather rapidly.

The crack willow is very similar to the weeping willow in most characters, except for its more coarsely serrate leaves. While most specimens of the crack willow lack decidedly pendulous branches, a few do possess this "weeping" character, making distinction from the weeping willow rather difficult.

The flowers are borne during April and May.

8. Salix babylonica L. Sp. Pl. 1017. 1753. *Fig. 8.*
Salix annularis Forbes, Salict. Woburn. 41. 1829.

Trees to 15 m tall, with a trunk diameter up to 0.5 m, the bark thick, rough, gray; twigs yellow or brown, glabrous, pendulous; leaves linear-lanceolate, long-acuminate at the apex, cuneate at the base, straight, serrulate, to 12 cm long, to 2 cm broad, sericeous when young, becoming glabrous or nearly so at maturity, yellow-green above, whitened below, the petiole to 1.2 cm long, glabrous, with glands at the upper end; stipules absent or up to 7 mm long; aments borne as the leaves begin to expand; staminate aments erect to ascending, slender, to 4 cm long, the flowers spirally arranged, each with 2 (−5) stamens, the filaments free, pilose near the base, the bract oblong, yellow, pubescent, caducous, the glands 1−2; pistillate aments erect to ascending, slender, to 3.5 cm long, the flowers spirally arranged, each with a glabrous ovary, the bract oblong, yellow, pubescent, caducous, the glands 1 (−2), cupular, the style less than 1 mm long; capsules ovoid-conical, 1−2 cm long, glabrous, sessile.

COMMON NAME: Weeping Willow.

HABITAT: Along roads, around homesteads.

RANGE: Native of Europe and Asia; escaped from cultivation in much of the eastern half of the United States.

ILLINOIS DISTRIBUTION: Scattered in most of the state.

Weeping willow is a commonly planted ornamental, popular primarily because of its pendulous ("weeping") branches and rapid manner of growth.

Mead reported this species in 1846 as *Salix annularis*, but this is considered to be the same species as *S. babylonica*. It may be confused with weeping forms of the crack willow, but

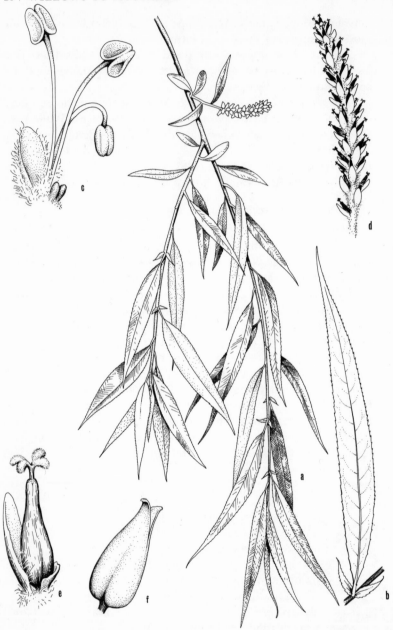

8. *Salix babylonica* (Weeping Willow). *a.* Fruiting branch, X⅜. *b.* Leaf, X¾.
c. Staminate flower, X15. *d.* Pistillate ament, X1½. *e.* Pistillate flower, X15.
f. Capsule, X15.

differs by its serrulate leaves, its smaller, sessile capsules, and its nonbrittle branches.

Other cultivated weeping willows, such as Thurlow's weeping willow (*S. elegantissima* Koch), are grown in Illinois, but have not been found as escapes. *Salix elegantissima* differs from *S. babylonica* by its broader leaves and stipules and longer aments.

Salix babylonica flowers during April and May.

9. Salix alba L. Sp. Pl. 1021. 1753.

Trees to 15 m tall, with a trunk diameter up to 0.5 m, the bark thick, rough, gray; twigs olive-green or brown or yellow, slender, pubescent or glabrous, not brittle; leaves lanceolate, acuminate at the apex, cuneate at the base, straight, serrulate, to 10 cm long, to 2.5 cm broad, sericeous or glabrous, more or less whitened on the lower surface, the petiole to 1 cm long, pubescent or glabrous, glandular at the upper end; stipules minute, caducous; aments borne as the leaves begin to expand; staminate aments erect to ascending, slender, to 3.5 cm long, the flowers spirally arranged, each flower with 2 (–3) stamens, the filaments free, pilose at the base, the bract oblong, yellow, sparsely pubescent, caducous, the glands 2; pistillate aments erect to ascending, slender, to 6 cm long, the flowers spirally arranged, each with a glabrous ovary, the bract oblong, yellow, sparsely pubescent, caducous, the gland 1, the style less than 1 mm long; capsules ovoid-conical, 3.0–4.5 mm long, glabrous, sessile or nearly so.

Three varieties are known from Illinois.

1. Leaves sericeous beneath _____ 9a. *S. alba* var. *alba*
1. Leaves glabrous or nearly so beneath.
   2. Branchlets brown _____ 9b. *S. alba* var. *calva*
   2. Branchlets yellow _____ 9c. *S. alba* var. *vitellina*

9a. Salix alba L. var. alba *Fig.* 9.

Leaves sericeous beneath.

COMMON NAME: White Willow.

HABITAT: Along streams.

RANGE: Native of Europe; escaped from cultivation in most of the eastern United States.

ILLINOIS DISTRIBUTION: Scattered throughout the state. During the early settlement of the United States, this plant was commonly planted because of its use in the mak-

9. *Salix alba* (White Willow). *a.* Fruiting branch, X½. *b.* Staminate flower, X10.
c. Pistillate flower, X10.

ing of charcoal. It is now sparingly found as an escape in the eastern United States.

The sericeous-leaved variety is more common in Illinois than either var. *calva* or var. *vitellina*.

The flowers open during April and May.

9b. Salix alba L. var. **calva** G. F. W. Mey.

Leaves glabrous or nearly so beneath; branchlets brown.

COMMON NAME: White Willow.

HABITAT: Along streams.

RANGE: Native of Europe; escaped from cultivation in the eastern United States.

ILLINOIS DISTRIBUTION: Known only from Jackson County.

The only Illinois specimen of this variety is a small tree about two meters tall.

9c. Salix alba L. var. **vitellina** (L.) Stokes, Bot. Mat. Med. 4:506. 1812.

Salix vitellina L. Sp. Pl. ed. 2, 1442. 1763.

Leaves glabrous or nearly so on the lower surface; branchlets yellow.

COMMON NAME: Golden Willow.

HABITAT: Along rivers and streams; disturbed moist soil.

RANGE: Native of Europe; occasionally escaped from cultivation in the eastern United States.

ILLINOIS DISTRIBUTION: Known from Champaign, Kankakee, and Vermilion counties.

Although the yellow branches of the golden willow give this a strikingly different appearance from the white willow, floral and fruiting similarities indicate it to be treated best as a variety of *Salix alba*. There is suspicion that this taxon may be a hybrid between S. *alba* and S. *fragilis*.

10. Salix interior Rowlee, Bull. Torrey Club 27:253. 1900.

Trees to 8 m tall, with shoots arising from vegetative buds on the roots, thus often forming dense colonies, with a trunk diameter up to 30 cm, the bark gray, rough at maturity; twigs brown to reddish-brown, slender, sericeous to nearly glabrous; leaves linear to linear-lanceolate, acuminate at the apex, cuneate at the base, straight, remotely glandular-denticulate, to 10 cm long, to 1 cm broad (sometimes somewhat larger on sprouts), sericeous when young, usually becoming sparsely pubescent at maturity, green on both sides, the petiole to 7 mm long, glabrous, eglandular; stipules minute or absent; aments borne as the leaves begin to expand; staminate aments erect to ascending, slender, sometimes branched, to 3 cm long, the flowers spirally arranged, each with 2 stamens, the filaments free, pilose in the lower half, the bract oblong, yellow, more or less pubescent, caducous, the gland 1; pistillate aments erect to ascending, slender, sometimes branched, to 7.5 cm long, the flowers spirally arranged, each with a sparsely sericeous or glabrous ovary, the bract narrowly oblong, yellow, pubescent, not persistent, with 1 gland, the style minute or absent; capsules narrowly conic-ovoid, 5–9 mm long, sparsely sericeous to glabrous, on pedicels twice as long as the gland.

Two forms occur in Illinois.

1. Leaves sparsely sericeous to nearly glabrous ----------------------- --10a. *S. interior* f. *interior*
1. Leaves densely sericeous --------------- 10b. *S. interior* f. *wheeleri*

10a. Salix interior Rowlee f. **interior** *Fig. 10.*
Salix longifolia Muhl. Neue Schrift. Ges. Nat. Fr. Berlin 4:238. 1803, non Lam. (1778).

Leaves sparsely sericeous to nearly glabrous.

COMMON NAME: Sandbar Willow.
HABITAT: Sandbars, banks of streams.
RANGE: Quebec to Alaska, south to Wyoming, Texas, and Maryland.
ILLINOIS DISTRIBUTION: Common throughout the state.
The sandbar willow is generally found in thickets or dense colonies in sandy, moist habitats throughout Illinois. The presence of vegetative buds among the roots is rare in the genus *Salix*.
Where it occurs on the sandy beaches of Lake Michigan, this willow grows with *Ammophila breviligulata*, *Artemisia caudata*. *Cakile edentula*, *Calamovilfa longifolia*, *Potentilla anserina*, and *Prunus pumila* (Swink, 1974).

Seedlings are reported by Lindsey et al. (1961) to have lobed leaves.

10. Salix interior (Sandbar Willow). *a.* Fruiting branch, X¾. *b.* Staminate flower, X5. *c.* Pistillate flowers, X5.

11. *Salix rigida* (Heart-leaved Willow).   *a.* Vegetative branch, X¾.   *b.* Staminate flower, X10.   *c.* Pistillate flower, X10.

Another unique feature of *Salix interior* is the occasional presence of branched aments.

The sandbar willow flowers in April and May.

10b. Salix interior Rowlee f. **wheeleri** (Rowlee) Rouleau, Nat. Canad. 71:268. 1944.

Salix interior Rowlee var. *wheeleri* Rowlee, Bull. Torrey Club 27:253. 1900.

Salix wheeleri (Rowlee) Rydb. ex Britt. Man. ed. 2, 1061. 1905.

Salix longifolia Muhl. var. *wheeleri* (Rowlee) Schneid. Bot. Gaz. 67:342. 1919.

Leaves densely sericeous.

COMMON NAME: Sandbar Willow.

HABITAT: Sandbars, banks of streams.

RANGE: Same as f. *interior*.

ILLINOIS DISTRIBUTION: Occasional in most parts of the state.

Forma *wheeleri* is distinguished by its permanently sericeous leaves. In addition, the leaves tend to be considerably shorter than those of f. *interior*. Exceptionally sericeous specimens look very different from typical *S. interior*.

Costello (1935) suggests that the sericeous leaves seem to be correlated with insect damage.

11. Salix rigida Muhl. Neue Schrift. Ges. Nat. Fr. Berlin 4:236. 1803. *Fig. 11.*

Salix cordata Muhl. Neue Schrift. Ges. Nat. Fr. Berlin 4:236. May, 1803, non Michx. (March, 1803).

Salix angustata Pursh, Fl. Am. Sept. 613. 1814.

Salix cordata Muhl. var. *rigida* (Muhl.) Carey ex Gray, Man. Bot. 427. 1848.

Salix cordata β *angustata* (Pursh) Anders. in DC. Prodr. 16: 252. 1868.

Shrubs to 3 m tall; twigs reddish-brown to yellowish, slender, pubescent when young, eventually becoming glabrous or nearly so; leaves oblong-lanceolate, acute to acuminate at the apex, rounded or subcordate at the base, serrulate, to 10 cm long, to 4 cm broad, densely sericeous when young, usually becoming glabrous or nearly so at maturity, usually reddish-purple when immature, the petioles to 1.5 cm long, pubescent, without glands at the upper end; stipules to 2 cm long, cordate, usually pubescent, persistent; aments borne just as the leaves begin to expand; staminate aments erect to ascending, to 3 cm long, the flowers spirally arranged, each with 2 stamens, the filaments free, glabrous, the bract obtuse, brown, pilose, not persistent, the gland 1; pistillate aments erect to ascending, to 5 cm long, the flowers spirally arranged, each with a glabrous ovary, the bract narrowly oblong, brown to black, pilose, subpersistent, the glands 2, the style less than 1 mm long; capsules narrowly ovoid, 4–5 mm long, glabrous, on pedicels usually less than twice as long as the gland.

COMMON NAME: Heart-leaved Willow.

HABITAT: Moist ground under a variety of conditions.

RANGE: Newfoundland to Ontario, south to Kansas, Mississippi, and North Carolina.

ILLINOIS DISTRIBUTION: Occasional throughout the state.

Salix rigida, *S. glaucophylloides* var. *glaucophylla*, and *S. syrticola* are three Illinois taxa in which the leaves are usually cordate or subcordate at the base. *Salix rigida* differs from *S. syrticola* by its narrower leaves, its eglandular-serrulate leaves, and its shorter pistillate aments and capsules. It differs from *S. glaucophylloides* var. *glaucophylla* by its narrower, serrate to crenate leaves.

Salix rigida also resembles *S. nigra* vegetatively, but differs by lacking petiolar glands and by never attaining the stature of a tree.

Much confusion has existed with respect to the nomenclature of this species. Muhlenberg described it in May of 1803 as *S. cordata*, but two months after Michaux had described a different North American species as *S. cordata*. Thus, with Muhlenberg's *S. cordata* unavailable for this species, the binomial *S. rigida* Muhl. becomes the earliest available.

Pursh had described a narrow-leaved variant as *S. angustata* in 1814 which Anderson later reduced to a variety.

In the present study, I am unable to separate satisfactorily var. *angustata* from the typical variety.

The heart-leaved willow flowers during April and May.

12. **Salix × myricoides** Muhl. Neue Schrift. Ges. Nat. Fr. Berlin 4:236. 1803. *Fig. 12.*

Salix cordata Muhl. var. *myricoides* (Muhl.) Carey ex Gray, Man. Bot. 427. 1848.

Shrub to 3 m tall; twigs brownish or yellowish, slender, usually

12. Salix × myricoides (Myrtle Willow). *a.* Vegetative branch, X½. *b.* Staminate inflorescence, X⅓. *c.* Staminate flower, X10. *d.* Pistillate flower, X10.

13. Salix glaucophylloides var. *glaucophylla* (Blue-leaf Willow). *a.* Vegetative branch, X½. *b.* Staminate flower, X6. *c.* Pistillate flower, X6. *d.* Fruiting raceme, X¾.

canescent-pubescent, brittle at the base; leaves lanceolate, long-acuminate at the apex, cuneate or subcuneate at the base, serrulate, to 10 cm long, to 3 cm broad, more or less sericeous beneath, green above, whitened below, the petiole to 1 cm long, more or less pubescent, without glands at the upper end; stipules very small, caducous; aments borne before the leaves begin to expand; staminate aments erect to ascending, slender, to 2.5 cm long, the flowers spirally arranged, each with 2 stamens, the filaments free, sparsely pubescent at the base, the bract narrowly oblong, dark brown, pilose, not persistent, the gland 1; pistillate aments erect to ascending, slender, to 4 cm long, the flowers spirally arranged, each with a thin sericeous ovary, the bract narrowly oblong, dark brown, pubescent, subpersistent, the gland 1, the style minute; capsules narrowly ovoid, 3–5 mm long, thinly sericeous, on pedicels usually less than twice as long as the gland.

COMMON NAME: Myrtle Willow.

HABITAT: Moist ground.

RANGE: Newfoundland to Ontario, south to Missouri and South Carolina.

ILLINOIS DISTRIBUTION: Known from Jackson County, but undoubtedly elsewhere.

Salix × myricoides is a reputed hybrid between S. rigida and S. sericea, although it appears to have more of the characteristics of S. sericea. However, it differs from S. sericea by its canescent twigs and thinly sericeous capsules, while S. sericea has glabrous or glabrate twigs and densely sericeous capsules.

13. **Salix glaucophylloides** Fern. var. **glaucophylla** (Bebb) Schneider, Journ. Arn. Arb. 1:157. 1920. *Fig. 13.*

Salix cordata Muhl. var. *glaucophylla* Bebb ex Babcock, The Lens 2:249. 1873.

Salix glaucophylla (Bebb) Bebb, Rep. Nat. Hist. Northwest. Univ. 1889:23. 1889, non Bess. (1822).

Salix glaucophylla (Bebb) Bebb var. *latifolia* Bebb, Rep. Nat. Hist. Northwest. Univ. 1889:23. 1889.

Salix glaucophylla (Bebb) Bebb var. *angustifolia* Bebb, Rep. Nat. Hist. Northwest. Univ. 1889:23. 1889.

Shrubs to 3 m tall; twigs yellowish or brown, slender, usually glabrous, shiny; leaves lanceolate to oblong to narrowly ovate, acute or short-acuminate at the apex, rounded to subcordate at the base, serrate or serrate-crenate, to 12 cm long, to 4.5 cm broad, somewhat pubescent when young, becoming glabrous or nearly so at maturity except for the puberulent midrib beneath, green above, glaucous beneath, the petiole to 1.0 (–1.2) cm long, mostly pubescent, without glands at the upper end; aments borne as the leaves begin to expand; stipules ovate, glandular-serrate, to 1 cm long, glaucous, persistent; staminate aments erect to ascending, to 5 cm long, the flowers spirally arranged, each with 2 stamens, the filaments free or nearly so, glabrous, the bract oblong, dark brown to black, villous, subpersistent, the gland 1; pistillate aments erect to ascending, to 10 cm long, loosely flowered, the flowers spirally arranged, each with a glabrous ovary, the bract oblong, dark brown to black, villous, subpersistent, the gland 1, the style 1.0–1.3 mm long; capsules conical-subulate, 4–9 mm long, glabrous, on pedicels much longer than the gland and usually longer than the bract.

COMMON NAME: Blue-leaf Willow.

HABITAT: Open sand, shores, marshes.

RANGE: Ontario south to Illinois, Indiana, and Ohio.

ILLINOIS DISTRIBUTION: Confined to the northern one-third of the state; also St. Clair County.

I am following Schneider and Fernald in considering our material as a variety of the more northern S. *glaucophylloides*. It differs from var. *glaucophylloides* by its longer pedicels and loosely flowered pistillate aments.

Swink (1974) reports this taxon as an occupant of open sand near Lake Michigan, along with *Cornus stolonifera* and *Salix syrticola*. He also reports it as occurring on calcareous pond shores, old sand pits, and on marsh borders.

The blue-leaf willow flowers during May.

A narrow-leaved variant, described as var. *angustifolia* by Bebb, was first collected by H. H. Babcock from Cook County. It is not recognized in this study.

14. Salix eriocephala Michx. Fl. Bor. Am. 2:225. 1803. *Fig. 14*.
Salix rigida Muhl. var. *vestita* Anderss. Mon. Sal. 159. 1867.
Salix missouriensis Bebb, Gard. & For. 8:373. 1895.

Trees to 15 m tall, with a trunk diameter to 0.5 m, the bark black; twigs brown to gray-black, slender, tomentose; leaves narrowly lanceolate to ovate-oblong, long-attenuate at the apex, cuneate or subcuneate at the base, serrulate, to 15 cm long, to 4 cm broad, pubescent beneath at maturity, at least on the nerves, dull green

14. Salix eriocephala (Willow).   *a.* Vegetative branch, X⅜.   *b.* Staminate flower,
X5.   *c.* Pistillate flower, X5.

15. *Salix syrticola* (Sand-dune Willow). *a*. Flowering and fruiting branch, X½. *b*. Staminate flower, X6. *c*. Pistillate flower, X6.

above, glaucous below, the petioles to 1 cm long, pubescent, without glands at the upper end; stipules reniform, to 1 cm long, persistent; aments borne before the leaves expand; staminate aments erect to ascending, to 5 cm long, the flowers spirally arranged, each with 2 stamens, the filaments free, pilose at the base, the bract oblong, dark brown to black, pubescent, subpersistent, the gland 1; pistillate aments erect to ascending, to 10 cm long, the flowers spirally arranged, each with a glabrous ovary, the bract oblong, dark brown to black, pubescent, subpersistent, the gland 1, the style about 1 mm long; capsules conical-subulate, to 1 cm long, glabrous, on pedicels much longer than the gland.

COMMON NAME: Willow.

HABITAT: Alluvial soil along streams.

RANGE: Minnesota to South Dakota, south to Nebraska and Kentucky.

ILLINOIS DISTRIBUTION: Confined to the southern one-third of the state; also Mason County.

This is the same species which Bebb described in 1895 as *S. missouriensis*, but Michaux's *S. eriocephala* is clearly identical and predates Bebb's binomial.

Some botanists consider *S. eriocephala* to be a variety of *S. rigida*. Jones et al. (1955) place *S. eriocephala* in synonymy under *S. discolor*, but there seems to be no good basis for this.

Salix eriocephala flowers during April.

15. Salix syrticola Fern. Rhodora 9:225. 1907. *Fig. 15.*

Shrub to 3 m tall; twigs gray, slender, tomentose or becoming glabrous; leaves oblong-ovate, acute or short-acuminate at the apex, cordate or broadly rounded at the base, glandular-serrate, to 9 cm long, to 6 cm broad, densely sericeous when young, becoming less pubescent or nearly glabrous at maturity, green above and below, the petiole to 1 cm long, pubescent, without glands at the upper end; stipules cordate, to 1.5 cm long, persistent; aments borne as the leaves begin to expand; staminate aments erect to ascending, to 4.5 cm long, the flowers spirally arranged, each with 2 stamens, the filaments free, glabrous, the bract oblong, brown, pubescent, subpersistent, the gland 1; pistillate aments erect to ascending, to 8 cm long, the flowers spirally arranged, each flower with a glabrous ovary, the bract oblong, brown, pubescent, subpersistent, the gland 1, the style 1.0–1.5 mm long; capsule conic-subulate, 5–7 mm long, glabrous, on pedicels more than twice as long as the gland.

COMMON NAME: Sand-dune Willow.

HABITAT: Low sand dunes.

RANGE: Western Ontario and Michigan, south to north-eastern Illinois and Indiana.

ILLINOIS DISTRIBUTION: Known from Cook and Lake counties.

This species has a very restricted range, occurring on low dunes around Lake Michigan. In addition to the two Illinois counties, it is reported by Argus (1964) from Manitowoc County, Wisconsin, Indiana Dunes State Park, Indiana, New Buffalo, Michigan, and Big Bay, Bruce Peninsula, Ontario.

Jones et al. (1955) and others do not distinguish S. *syrticola* from S. *cordata* Michx. or S. *adenophylla* Hook., but I believe enough significant differences exist to justify specific status for S. *syrticola*.

The flowers of this species appear in June.

16. Salix bebbiana Sarg. Gard. & For. 8:463. 1895. *Fig. 16.*

Salix rostrata Richardson, Frankl. Journ. App. 753. 1823, non Thuill. (1799).

Salix starkeana Willd. ssp. *bebbiana* (Sarg.) Youngberg, Rhodora 72:549. 1970.

Shrubs or small trees to 6 m tall; twigs reddish-brown to dark brown, tomentose at least when young; leaves oblanceolate to ovate-oblong, acute at the apex, cuneate to rounded at the base, entire or irregularly glandular-crenate, to 7.5 cm long, to 3.5 cm broad, pilose to tomentose when young, becoming sparsely pubescent at maturity, dull green above, glaucous and rugose beneath, the petiole to 1 cm long, pubescent, without glands at the tip; stipules small, caducous; aments borne as the leaves begin to expand; staminate aments erect to ascending, to 2.5 cm long, the flowers spirally arranged, each with 2 stamens, the filaments free, pilose at the base, the bract lanceolate, yellowish, pubescent, subpersistent, the gland 1; pistillate aments erect to spreading, to 5 cm long, loosely flowered, the flowers spirally arranged, each flower with a sericeous ovary, the bract lanceolate, yellow, pubescent, subpersistent, the gland 1, the style minute or absent; capsules conical-subulate, 5–9 mm long, pubescent, on pedicels many times longer than the gland.

16. *Salix bebbiana* (Bebb Willow). *a.* Vegetative branch, X½. *b.* Leaf, X1. *c.* Staminate flower, X6. *d.* Pistillate flower, X6. *e.* Fruiting branch, X1.

17. *Salix pedicellaris* (Bog Willow). *a.* Habit, in fruit, X¼. *b.* Staminate inflorescence, X¾. *c.* Staminate flower, X6. *d.* Pistillate flower, X6.

COMMON NAME: Bebb Willow.

HABITAT: Boggy soils.

RANGE: Newfoundland to Alberta, south to Iowa, northern Illinois, Ohio, Pennsylvania, and Maryland.

ILLINOIS DISTRIBUTION: Confined to the northern one-fourth of Illinois.

Salix bebbiana is named for Michael Shuck Bebb, an authority on *Salix*, who lived at Fountaindale southwest of Rockford during the last half of the nineteenth century.

Considerable variation exists in the degree of pubescence on the leaves and the amount of serration on the margins. It often happens that the lowermost leaves on the twigs are entire, while the uppermost leaves are crenate.

Youngberg (1970) considers S. *bebbiana* to be a subspecies of the European S. *starkeana*.

Swink (1974) reports that *Salix bebbiana* occurs in the shrub zone of bogs where there has been disturbance.

The trinomial applied to this taxon by Youngberg (1970) is illegitimate.

The flowers are borne in May.

17. Salix pedicellaris Pursh, Fl. Am. Sept. 611. 1814. *Fig. 17.*
Salix pedicellaris Pursh var. *hypoglauca* Fern. Rhodora 11:161. 1909.

Shrubs to 0.75 m tall, sometimes decumbent and rooting at the nodes; twigs yellow to reddish-brown to gray, glabrous; leaves oblong to oblong-obovate, obtuse to acute at the apex, rounded to subcuneate at the base, entire and somewhat revolute, to 7.5 cm long, to 2.5 cm broad, glabrous, dark green above, mostly glaucous beneath, the petiole to 6 mm long, glabrous, without glands at the upper end; stipules absent; aments borne as the leaves begin to expand; staminate aments erect to ascending, to 2 cm long, the flowers spirally arranged, each with 2 stamens, the filaments free or nearly so, glabrous, the bract pale yellow, pubescent, subpersistent, the gland 1; pistillate aments erect to ascending, to 3 cm long, the flowers spirally arranged, each with a glabrous ovary, with the bract pale yellow, pubescent, subpersistent, the gland 1; capsule oblongoid-conic, 4–8 mm long, glabrous, on pedicels much longer than the glands.

COMMON NAME: Bog Willow.

HABITAT: Bogs.

RANGE: Newfoundland to British Columbia, south to Oregon, Iowa, and New Jersey.

ILLINOIS DISTRIBUTION: Confined to the northern one-third of the state.

Salix pedicellaris is the only willow in Illinois with entire, revolute leaves completely glabrous beneath.

All Illinois specimens are somewhat glaucous on the lower surface and have been segregated by some workers as var. *hypoglauca* Fern. Argus (1964) gives evidence that var. *hypoglauca* is not sufficiently distinct to merit varietal status.

Swink (1974) reports that this species is an associate of *Salix candida* and *Betula pumila*.

The bog willow flowers during April and May.

18. Salix discolor Muhl. Neue Schrift. Ges. Nat. Fr. Berlin 4:234. 1803. *Fig. 18.*

Salix prinoides Pursh, Fl. Am. Sept. 613. 1814.

Salix sensitiva Barratt, Salic. Am. no. 8. 1840.

Salix discolor Muhl. var. *sensitiva* (Barratt) Bebb, Trans. Ill. State Agr. Soc. 3:587. 1859.

Salix discolor Muhl. var. *latifolia* Anderss. Sv. Vet. Akad. Handl. 6:84. 1867.

Salix imponens Gandoger, Fl. Europ. 21:167. 1890.

Shrubs or small trees to 6 m tall; twigs brown, pubescent, sometimes becoming glabrous by the second year; leaves lanceolate to elliptic to obovate, acute at the apex, rounded to subcuneate at the base, serrate to crenate, to 10 cm long, to 3 cm broad, ferruginous-pilose when young, usually becoming glabrous or nearly so, dark green above, glaucous below, the petiole to 1.5 cm long, usually pubescent, without glands at the upper end; stipules prominent on sprouts; aments borne before the leaves begin to expand, sessile or nearly so; staminate aments erect to ascending, to 5 cm long, the flowers spirally arranged, each with 2 stamens, the filaments free, glabrous or pilose near the base, the bract brown or black, conspicuously white-villous, persistent, the gland 1; pistillate aments erect to ascending, to 7 cm long, densely flowered, the flowers spirally arranged, each with a densely sericeous ovary, the bract brown or black, white-villous, subpersistent, the lowest ones tawny and usually not subtending a flower, the gland 1, the style up to 1 mm long;

18. Salix discolor (Pussy Willow). *a.* Vegetative branch, X⅜. *b.* Staminate flower, X5. *c.* Pistillate inflorescence, X¼. *d.* Pistillate flower, X5. *e.* Fruiting raceme, X¾.

capsules oblongoid-conic, 6–12 mm long, puberulent, the pedicel about twice as long as the gland.

COMMON NAME: Pussy Willow.

HABITAT: Marshes, swamps.

RANGE: Labrador to Alberta, south to Montana, Missouri, and Maryland.

ILLINOIS DISTRIBUTION: Occasional in the northern four-fifths of the state.

Salix discolor is variable in the degree and persistence of the pubescence of twigs and leaves. Leaves which remain permanently pubescent have been designated var. *latifolia* Anderss., but Argus (1964) speculates that these latter plants may be hybrids between *Salix discolor* Muhl. and *Salix humilis* Marsh. At any rate, the pubescent forms are not very well defined and are not recognized as distinct in this work.

In northeastern Illinois, *S. discolor* frequently is associated with other willows such as *S. bebbiana*, *S. petiolaris*, and *S. interior*.

Feldman's (1942) report of *S. prinoides* from Champaign County is actually this species.

The pussy willow sold by most florists is not this species but *S. caprea* L., the goat willow.

Salix discolor flowers from March to May.

19. Salix humilis Marsh. Arb. Am. 140. 1785.

Shrubs to 3 m tall; twigs yellow to brown, pubescent to nearly glabrous; leaves linear-lanceolate to oblanceolate to elliptic, acute to short-acuminate at the apex, subcuneate to cuneate at the base, undulate to crenate or sometimes nearly entire, revolute, to 10 cm long, to 3 cm broad, densely pubescent when young, sometimes becoming glabrate at maturity, green or grayish and dull above, whitened below, the petiole to 1.5 cm long, pubescent, without glands at the upper end; stipules linear-lanceolate, to 1 cm long, caducous; aments borne before the leaves begin to expand, sessile or nearly so; staminate aments erect to ascending, to 3 cm long, the flowers spirally arranged, each with 2 stamens, the filaments free, glabrous, the bract brown or black, villous, subpersistent, the gland 1; pistillate aments erect to ascending, to 5 cm long, the flowers spirally arranged, each with a sericeous ovary, the bract brown or black, villous, subpersistent, the lowest ones greenish and not subtending a flower, the gland 1, the style less than 1 mm long; capsules narrowly conic, 4–9 mm long, densely pubescent to tomentulose, the pedicel several times longer than the gland.

Three varieties may be recognized in Illinois.

1. Staminate aments 1 cm thick or thicker; leaves (4–) 5–10 cm long, on petioles 3 mm long or longer.
1. Staminate aments less than 1 cm thick; leaves 1.5–5.0 cm long, on petioles less than 3 mm long _____ 19c. S. humilis var. microphylla
2. Leaves densely soft-pubescent beneath _____ _____ 19a. S. humilis var. humilis
2. Leaves glabrous or sparsely puberulent beneath _____ _____ 19b. S. humilis var. hyporhysa

19a. Salix humilis Marsh. var. **humilis** *Fig. 19.*

Staminate aments 1 cm thick or thicker; leaves (4–) 5–10 cm long, on petioles 3 mm long or longer; leaves densely soft-pubescent beneath; twigs 2 mm or more thick.

COMMON NAME: Prairie Willow.

HABITAT: Prairies, sandy soil.

RANGE: Newfoundland to Minnesota, south to Kansas, Louisiana, and North Carolina.

ILLINOIS DISTRIBUTION: Scattered throughout the state.

This permanently pubescent, large-leaved variety of the prairie willow is seemingly a little more common in Illinois than the other two varieties, particularly in the northern counties.

The leaves tend to enlarge after the capsules have shed their seeds.

The flowers are borne from March to May.

19b. Salix humilis Marsh. var. **hyporhysa** Fern. Rhodora 48:45. 1946. *Fig. 20.*

Salix humilis Marsh. var. *longifolia* f. *rigidiuscula* Anders. Sv. Vet. Akad. Handl. 6:111. 1867.

Salix humilis Marsh. var. *rigidiuscula* (Anders.) Robins. & Fern. in Gray, Man. Bot. ed. 7, 326. 1908.

Staminate aments 1 cm thick or thicker; leaves (4–) 5–10 cm long, on petioles 3 mm long or longer; leaves glabrous or sparsely puberulent beneath; twigs 2 mm or more thick.

19. *Salix humilis* var. *humilis* (Prairie Willow). *a.* Vegetative branch, X½. *b.* Staminate flower, X7½. *c.* Pistillate inflorescence, X¼. *d.* Pistillate flower, X7½.

20. *Salix humilis* var. *hyporhysa* (Smooth Prairie Willow). *a.* Vegetative branch, X½. *b.* Leaf, X1. *c.* Staminate inflorescence, X¼. *d.* Staminate flower, X7½. *e.* Pistillate flower, X7½. *f.* Fruiting branch, X¼.

COMMON NAME: Smooth Prairie Willow.

HABITAT: Prairies.

RANGE: Connecticut to Wisconsin and Iowa, south to eastern Texas and Florida.

ILLINOIS DISTRIBUTION: Scattered throughout the state.

Salix humilis var. *hyporhysa* differs from var. *humilis* by its less pubescent leaves. Since some intergradation takes place in Illinois, there may be little justification to recognize it as distinct.

The flowers appear from March to May.

19c. Salix humilis Marsh. var. **microphylla** (Anderss.) Fern. Rhodora 48:46. 1946. *Fig. 21.*

Salix tristis Ait. Hort. Kew. 3:393. 1789.

Salix tristis Ait. var. *microphylla* Anderss. Proc. Am. Acad. Arts & Sci. 4:67. 1858.

Salix humilis Marsh. var. *tristis* (Ait.) Griggs, Proc. Ohio Acad. 4:301. 1905.

Staminate aments less than 1 cm thick; leaves 1.5–5.0 cm long, on petioles less than 3 mm long.

COMMON NAME: Sage Willow.

HABITAT: Prairies.

RANGE: Maine to Minnesota, south to Oklahoma and Florida.

ILLINOIS DISTRIBUTION: Scattered in the northern half of the state.

The sage willow is considered by some workers to be a distinct species, while others believe it does not even merit varietal status. That it is generally distinct in its appearance cannot be denied. Argus (1964), without experimental evidence, suggests that it may be an ecological variant of *S. humilis*.

The flowers are borne in April and May.

20. Salix petiolaris Sm. Trans. Linn. Soc. 6:122. 1802. *Fig. 22.*

Salix petiolaris Sm. var. *gracilis* Anderss. in DC. Prodr. 16:235. 1868.

Salix gracilis Anderss. var. *textoris* Fern. Rhodora 48:46. 1946.

Shrubs to 3 m tall; twigs dark brown to yellow-green, pubescent or nearly glabrous; leaves linear to lanceolate, acuminate at the apex,

21. *Salix humilis* var. *microphylla* (Sage Willow).   *a.* Vegetative branch, X½.   *b.* Pistillate inflorescence, X¼.   *c.* Staminate flower, X10.   *d.* Pistillate flower, X10.

22. *Salix petiolaris* (Petioled Willow). *a.* Vegetative branch, X⅓. *b.* Staminate flower, X7½. *c.* Pistillate flower, X7½. *d.* Fruiting branch, X1.

cuneate at the base, serrate, denticulate, or subentire, to 12 cm long, to 2 cm broad, sericeous when young, becoming glabrous or sparsely sericeous at maturity, green above, glaucous beneath, the petiole to 1 cm long, pubescent, without glands at the upper end; stipules minute, caducous; aments borne as the leaves begin to expand; staminate aments erect to ascending, to 2 cm long, the flowers spirally arranged, each with 2 stamens, the filaments free, glabrous or pilose at the base, the bract oblong, brown, pubescent, subpersistent, the gland 1; pistillate aments erect to ascending, to 3.5 cm long, the flowers spirally arranged, each with a sericeous ovary, the bract oblong, brown, pubescent, subpersistent, the gland 1, the style minute or absent; capsule oblongoid-conical, 5–8 mm long, sericeous, the pedicels several times longer than the gland.

COMMON NAME: Petioled Willow.

HABITAT: Low prairies, marshes, bogs.

RANGE: Quebec to Manitoba, south to Nebraska, northern Illinois, and New Jersey.

ILLINOIS DISTRIBUTION: Confined to the northern one-third of the state.

The nomenclature regarding this taxon is confusing. Fernald (1946) calls it *S. gracilis* Anderss. var. *textoris* Fern., arguing that *S. petiolaris* Sm. is an entirely different European species. Ball (1948), on the other hand, believes that *S. petiolaris* Sm. was described from an American species cultivated in a European garden.

There is no doubt that *S. petiolaris* is related to *S. sericea*, with *S. petiolaris* differing primarily by its nearly glabrous leaves, pointed capsules, and later time for flowering.

The flowers of *S. petiolaris* are borne from April to June.

21. Salix sericea Marsh. Arb. Am. 140. 1785. *Fig. 23*.

Shrubs to 3 m tall; twigs brown, slender, glabrous or sparsely pubescent; leaves narrowly lanceolate, acuminate at the apex, cuneate at the base, glandular-serrulate, to 10 cm long, to 2.5 cm broad, dark green and glabrous or pubescent above, with appressed silky hairs beneath, the petioles to 1 cm long, more or less pubescent, without glands at the upper end; stipules (on sprouts) lanceolate, caducous; aments borne before the leaves begin to expand; staminate aments erect to ascending, to 2.5 cm long, the flowers spirally arranged, each with 2 stamens, the filaments free, pilose near the base, the bract oblong, dark brown to black, pubescent, persistent,

23. *Salix sericea* (Silky Willow).   *a.* Vegetative branch, X½.   *b.* Staminate flower, X7½.   *c.* Pistillate flower, X7½.   *d.* Fruiting branch, X½.

the gland 1; pistillate aments erect to ascending, to 2.5 cm long, the flowers spirally arranged, each with a sericeous ovary, the bract oblong, dark brown to black, pubescent, persistent, the gland 1, the style absent or minute; capsules ovoid-oblongoid, 3–5 mm long, sericeous, on pedicels several times longer than the gland.

COMMON NAME: Silky Willow.

HABITAT: Low ground, along streams, in bogs.

RANGE: Quebec to Wisconsin, south to Missouri and South Carolina.

ILLINOIS DISTRIBUTION: Occasional and scattered throughout the state.

The silky willow is a handsome shrub with its brown branches and silvery-silky leaves. It seems to be most common along streams.

The flowers are borne from late March until the middle of May.

22. Salix caprea L. Sp. Pl. 1020. 1753. *Fig. 24.*

Small trees or shrubs to 8 m tall, the bark gray; twigs brown, shiny, slender, pubescent at first, soon becoming glabrous; leaves elliptic to broadly ovate, acute at the apex, subcordate, rounded, or subcuneate at the base, more or less dentate to coarsely undulate, to 10 cm long, to 6 cm broad, softly hairy at first, becoming glabrous or nearly so, the petioles to 2 cm long, pubescent, without glands at the upper end; stipules (on sprouts) subpersistent; aments borne before the leaves begin to expand; staminate aments erect to ascending, thick, to 4 cm long, the flowers spirally arranged, each with 2 stamens, the filaments free, slightly pubescent at the base, the bract elliptic, nearly black, pubescent, subpersistent, the gland 1; pistillate aments erect or ascending, thick, to 10 cm long, the flowers spirally arranged, each with a pubescent ovary, the bract elliptic, nearly black, pubescent, subpersistent, the gland 1, the style minute; capsules ovoid-conical, pubescent, 6–8 mm long, on pedicels several times longer than the gland.

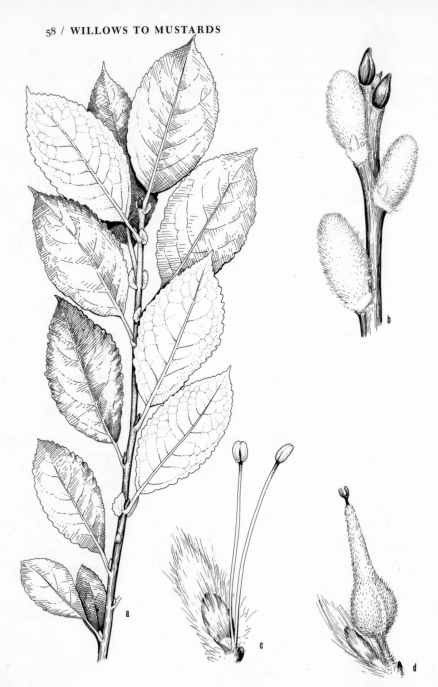

24. Salix caprea (Goat Willow). *a.* Vegetative branch, X½. *b.* Staminate inflorescence, X¾. *c.* Staminate flower, X6. *d.* Pistillate flower, X6.

COMMON NAME: Goat Willow.

HABITAT: Remnant around farm houses.

RANGE: Native of Europe; frequently planted but rarely escaped from cultivation.

ILLINOIS DISTRIBUTION: Known from Jackson, Lake, and Wabash counties.

The goat willow, with its large, "fuzzy" pistillate catkins, is generally the pussy willow of most florists. It flowers very early in March or April, well before the leaves begin to unfold.

Old specimens may attain a height of at least eight meters.

23. Salix candida Fluegge ex Willd. Sp. Pl. 4:708. 1806. *Fig. 25.*

Shrubs to 3 m tall; twigs yellow to brown, stout, whitish-tomentose when young, becoming glabrous or nearly so; leaves linear to oblong, to 10 cm long, acute at the apex, cuneate at the base, entire to undulate, revolute, densely whitish-tomentose when young, remaining whitish-tomentose or becoming sparsely tomentose at maturity, the petiole to 1 cm long, pubescent, without glands at the upper end; stipules lanceolate, tomentose, subpersistent; aments borne before the leaves begin to expand; staminate aments erect to ascending, dense, to 1.5 cm long, the flowers spirally arranged, each with 2 stamens, the filaments free, glabrous, the bract pale brown, pubescent, subpersistent, the gland 1; pistillate aments erect to ascending, dense, to 5 cm long, the flowers spirally arranged, each with a pubescent ovary, the bract pale brown, pubescent, subpersistent, the gland 1, the style about 1 mm long; capsules ovoid-conic, 4–6 mm long, densely tomentose, the pedicel about twice as long as the gland.

COMMON NAME: Hoary Willow.

HABITAT: Bogs.

RANGE: Labrador to British Columbia, south to Colorado, northern Illinois, and New Jersey.

ILLINOIS DISTRIBUTION: Confined to the northern one-third of the state.

The hoary willow is a species of bogs and moist prairies, where it associates with such species as *Betula pumila*, *Larix laricina*, *Rhus vernix*, and *Vaccinium macrocarpon*. This species flowers during April and May.

25. *Salix candida* (Hoary Willow).   *a.* Vegetative branch, X½.   *b.* Staminate in-florescence, X½.   *c.* Staminate flower, X7½.   *d, e.* Pistillate flowers, X7½.

24. Salix purpurea L. Sp. Pl. 1017. 1753. *Fig. 26.*

Shrubs to 4 m tall; twigs yellow-green, brown, or purple, slender, glabrous; leaves subopposite, linear to spatulate, acute to acuminate at the apex, more or less rounded at the base, entire or serrulate, to 10 cm long, to 1.5 cm broad, glabrous, often purple-tinged, more or less glaucous, the petioles to 0.5 cm long, glabrous, without glands at the tip; stipules absent; aments flowering before the leaves begin to expand; staminate aments erect to ascending, slender, to 3 cm long, the flowers spirally arranged, each with 2 stamens, the filaments united, pubescent near the base, the bract obovate, black-tipped, glabrous, subpersistent, the glands 2; pistillate aments erect to ascending, slender, to 3 cm long, the flowers spirally arranged, each with a pubescent ovary, the bract obovate, black-tipped, glabrous, subpersistent, the gland 1, the style minute; capsules broadly ovoid, 2–3 mm long, pubescent, sessile.

COMMON NAME: Purple Osier.

HABITAT: "Ditches," according to Pepoon (1927).

RANGE: Native of Europe; occasionally cultivated but rarely escaped in Illinois.

ILLINOIS DISTRIBUTION: Reported by Pepoon (1927) from northwest of Chicago, presumably Cook County; also Lake County.

The purple osier differs from most other willows in Illinois in several of its characteristics. Its leaves appear to be suboppositely arranged, its two filaments are united as one, and its capsules are sessile.

The leaves often have a peculiar purplish but yet silvery-gray color. The young twigs frequently are purple.

This species was reported by Pepoon (1927) "to be increasing in numbers" and "becoming common particularly northwest of Chicago."

The flowers are borne in April and May.

25. Salix × subsericea (Anderss.) Schneid. Ill. Handb. Laub-holzk. 1:65. 1904. *Fig. 27.*

Salix petiolaris Sm. var. *subsericea* Anderss. in DC. Prodr. 16(2):234. 1864.

Shrubs to 3 m tall; twigs brown, slender, usually somewhat seri-ceous when young, usually becoming less pubescent by maturity; leaves narrowly lanceolate, acuminate at the apex, cuneate at the base, entire to undulate to glandular-serrulate, to 12 cm long, to 3

26. *Salix purpurea* (Purple Osier).   *a.* Vegetative branch, X½.   *b.* Staminate flower, X10.   *c.* Pistillate flower, X10.

27. *Salix* × *subsericea* (Hybrid Willow). *a*. Vegetative branch, X⅜. *b*. Leaf, X¾. *c*. Staminate flower, X6. *d*. Pistillate flower, X6.

cm broad, puberulent at least on the lower surface, the petioles up to 1 cm long, more or less pubescent, without glands at the upper end; stipules (on sprouts) lanceolate, caducous; aments borne as the leaves expand; staminate aments erect to ascending, to 2 cm long, the flowers spirally arranged, each with 2 stamens, the filaments free, usually pilose near the base, the bract oblong, brown with a black tip, pubescent, subpersistent, the gland 1; pistillate aments erect to ascending, to 3 cm long, the flowers spirally arranged, each with a sericeous ovary, the bract oblong, brown with a black tip, pubescent, subpersistent, the gland 1, the style minute or absent; capsule ovoid, up to 6 mm long, sericeous, on pedicels several times longer than the gland.

COMMON NAME: Hybrid Willow.
HABITAT: Low ground.
RANGE: Quebec to Manitoba, south to Illinois and Virginia.
ILLINOIS DISTRIBUTION: Known only from Cook and Kankakee counties.
This taxon is a reputed hybrid between *Salix sericea* and *S. petiolaris* and it shares characteristics of both species. It is intermediate between these two species in degree of pubescence. It differs from both by having brown bracts with a black tip. It was originally described by Andersson as a variety of *S. petiolaris*.

The specimens seen from Cook and Kankakee counties have been identified by Carlton R. Ball as this hybrid.

The flowers are borne in late April and May.

2. *Populus* L.—Poplar

Dioecious trees; buds with several scales; leaves alternate, petiolate, with minute, caducous stipules; flowers unisexual, without a perianth; staminate flowers in dense, pendulous aments, from a cyathiform disc, each flower subtended by a dissected bract, with 4–60 free stamens; pistillate flowers in erect, spreading, or pendulous aments, from a cyathiform disc, each flower subtended by a dissected bract, with one pistil, the ovary superior, the stigmas 2–4, entire or 4-lobed; capsule 2- to 4-valved, with several comose seeds.

The leaves nearly as broad as long and the several-scaled buds distinguish *Populus* from *Salix*.

There are about thirty species of *Populus*, all native to the northern hemisphere. Several species are native to the western United States.

A number of species of *Populus* are frequently cultivated as ornamentals. These include the white poplar, or abele (*P. alba* L.), the European aspen (*P. tremula* L.), the black poplar (*P. nigra* L.) and its variety, the Lombardy poplar (var. *italica* DuRoi), the balsam poplar (*P. balsamifera* L.), and the balm-of-Gilead (*P.* × *gileadensis* Rouleau).

KEY TO THE TAXA OF *Populus* IN ILLINOIS

1. Petioles flat _____ 2
1. Petioles round _____ 6
  2. Leaves triangular-ovate to rhombic _____ 3
  2. Leaves ovate to suborbicular _____ 4
3. Leaves triangular-ovate; tree with a broad crown, the branches spreading _____1. *P. deltoides*
3. Leaves rhombic; tree with a narrow crown, the branches strongly ascending _____2. *P. nigra* var. *italica*
  4. Margin of leaf dentate, with 5–25 teeth (averaging 10–20); buds pubescent _____5
  4. Margin of leaf finely crenate, with 20 or more teeth (averaging 31); buds glabrous or nearly so _____ 5. *P. tremuloides*
5. Margin of leaf with 5–15 teeth (averaging 10); petiole 5–10 cm long (averaging 7 cm) _____ 3. *P. grandidentata*
5. Margin of leaf with 12–25 teeth (averaging 20); petiole 3–6 cm long (averaging 5.5 cm) _____ 4. *P.* × *smithii*
  6. Leaves covered with a white felt on the lower surface _____
  _____6. *P. alba*
  6. Leaves glabrous or variously pubescent beneath, but not covered with a white felt on the lower surface _____ 7
7. Leaves rounded or truncate at base, if cordate, the buds heavily resinous _____8
7. Leaves cordate at base; buds not heavily resinous _____
  _____ 10. *P. heterophylla*
  8. Leaves sinuate-dentate; buds to 6 mm long, not resinous _____
  _____7. *P. canescens*
  8. Leaves serrulate or crenulate; buds to 25 mm long, resinous _____
  _____9
9. Twigs usually glabrous; leaves glabrous, or with a few hairs on the midvein beneath _____ 8. *P. balsamifera*
9. Twigs usually pubescent; leaves pubescent beneath _____
  _____9. *P.* × *gileadensis*

1. Populus deltoides Marsh. Arb. Am. 106. 1785. *Fig. 28.*

Populus monilifera Ait. Hort. Kew. 3:406. 1789.

Populus nigra L. β *virginiana* Castiglioni Viagg. Stati Uniti 2:334. 1790.

Populus angulata Ait. var. *missouriensis* Henry in Elwes & Henry, Trees Gt. Brit. & Irel. 7:1811. 1913.

Populus deltoidea var. *missouriense* (Henry) Henry, Gard. Chron. ser. 3, 56:46. 1914.

Populus balsamifera L. var. *virginiana* (Castiglioni) Sarg. Journ. Arn. Arb. 1:63. 1919.

Populus balsamifera var. *missouriensis* (Henry) Rehder, Man. Cult. Trees & Shrubs 92. 1927.

Trees to 35 m tall, with a trunk diameter up to 2.5 m, the bark smooth and gray when young, becoming furrowed at maturity; leaves triangular-ovate, short-acuminate at the apex, truncate or slightly cordate at the base, ciliolate, to 12 cm long, nearly as broad or slightly broader, with 2–3 basal glands, more or less glabrous on both surfaces, the petiole flat, up to 10 cm long, glabrous; stipules small, caducous; staminate aments short-stalked, pendulous, stout, densely flowered, to 5 cm long, borne as the leaves begin to expand, the disk of each flower broad, asymmetrical, with about 60 stamens inserted on the disk, the filaments short, distinct, the bracts light brown, scarious, glabrous, fimbriate; pistillate aments short-stalked, pendulous or spreading, sparsely flowered, to 8 cm long, borne as the leaves begin to expand, the disk of each flower broad, crenulate, with one glabrous pistil, with 3–4 laciniate-lobed stigmas, the bracts light brown, scarious, glabrous, fimbriate; fruiting aments to 12 cm long, the capsules ovoid to ellipsoid, acute, 3- to 4-valved, to 10 mm long, glabrous, on glabrous pedicels to 6 mm long, the light brown seeds oblongoid-obovoid, to 2 mm long, with a cottony coma.

COMMON NAME: Cottonwood.

HABITAT: Along rivers and streams.

RANGE: New Brunswick to Alberta, south to south-central Texas, east to northern Florida.

ILLINOIS DISTRIBUTION: Common throughout the state.

The cottonwood is one of the most common trees in moist situations in Illinois. It is extremely fast-growing, sometimes attaining heights in excess of 30 meters in about one hundred years. The trees are relatively short-lived, however. The largest cottonwood in existence today in Illinois

28. *Populus deltoides* (Cottonwood). *a*. Branch with leaves and fruits, X½. *b*. Staminate ament, X1½. *c*. Staminate flower, X7½. *d*. Winter twig with pistillate ament, X1¼. *e*. Pistillate flower, X7½.

occurs about one mile from Gebhard Woods State Park, Grundy County. It has a circumference of nine feet, one inch at a distance 4½ feet above the ground.

Specimens with the leaves somewhat broader than long have been called var. *missouriense* (Henry) Henry, but I do not believe this variation is deserving of nomenclatural recognition.

The beadlike clusters of unopened capsules caused Aiton to describe the cottonwood as *P. monilifera*, but this binomial is antedated by Marshall's *P. deltoides*.

The primary uses for cottonwood are pulp for papermaking and for the manufacture of packing boxes. The wood, which is relatively light and tends to warp, is not valuable as a timber tree, although pioneers used it in construction of their villages.

The cottonwood flowers from March to May.

2. **Populus nigra** L. var. **italica** Muenchh. Haustvat. 5:230. 1770. *Fig. 29.*
Populus italica (Muenchh.) Moench, Baume Weiss. 79. 1785.
Populus dilatata Ait. Hort. Kew. 3:406. 1789.

Trees to 25 m tall, with a trunk diameter up to 1.5 m, with a narrow crown, the bark dark, furrowed, the twigs slender, brown, the buds acute, appressed close to the twig, to 8 mm long; leaves rhombic, acuminate at the apex, truncate at the base, crenate-serrate, only sparsely so across the base, ciliolate, to 8 cm long, about as broad or slightly broader, without basal glands, pubescent at first, more or less glabrous on both surfaces at maturity, the petiole flat, up to 8 cm long, glabrous; stipules small, caducous; staminate aments short-stalked, pendulous, densely flowered, to 5 cm long, borne as the leaves begin to expand, the disk of each flower broad, asymmetrical, with about 20 stamens inserted on the disk, the filaments short, distinct, the bracts brown, scarious, glabrous, fimbriate; pistillate aments unknown; fruits unknown.

COMMON NAME: Lombardy Poplar.
HABITAT: Along roads and around homesteads.
RANGE: First discovered in Europe; rarely escaped from cultivation.
ILLINOIS DISTRIBUTION: Scattered in Illinois.
The Lombardy poplar is considered to be a sterile "sport" of the European black cottonwood (*P. nigra* L.). It reproduces only by sprouts.
Although it is commonly planted, particularly as a fast-

29. *Populus nigra* var. *italica* (Lombardy Poplar). *a.* Vegetative branch, X½.

growing hedge, its short life and susceptibility to a canker disease make it undesirable.

The staminate flowers are borne from March to May.

3. Populus grandidentata Michx. Fl. Bor. Am. 2:243. 1803.
Fig. 30.

Populus grandidentata Michx. f. *meridionalis* Tidestrom, Rhodora 16:205. 1914.

Trees to 20 m tall, with a trunk diameter up to 0.5 m, the crown

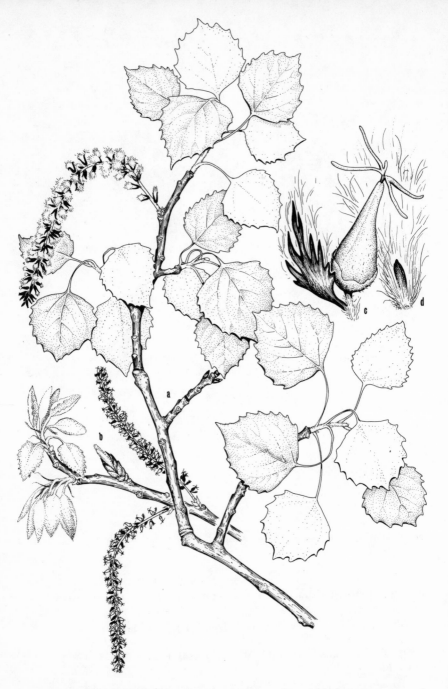

30. Populus grandidentata (Large-toothed Aspen). *a.* Branch with leaves and fruits, X¼. *b.* Branch with pistillate aments, X½. *c.* Pistillate flower, X6. *d.* Seed, X7½.

narrowly round-topped, the bark smooth, gray, when old becoming dark brown and divided into roughened, broad, flat ridges, the twigs rather stout, dark orange-brown, tomentose at first, becoming glabrous and shiny at maturity, with scattered orange lenticels, the buds ovoid, acute, light brown, pubescent, to 4 mm long; leaves broadly ovate to nearly orbicular, short-acuminate at the apex, cuneate to rounded at the base, dentate with 5–15 teeth (averaging 10), to 12 cm long, about as broad, densely white-tomentose when young, becoming sparsely pubescent to nearly glabrous when mature, dark green above, paler beneath, the petiole flat, up to 10 cm long, densely tomentose when young, becoming glabrous; stipules small, caducous; staminate aments pendulous, densely flowered, to 5 cm long, borne as the leaves begin to expand, the disk of each flower asymmetrical, entire, with 6–12 stamens, the filaments short, distinct, the bracts pale brown, scarious, pubescent, lobed; pistillate aments pendulous or spreading, sparsely flowered, to 10 cm long, borne as the leaves begin to expand, the disk of each flower asymmetrical, crenulate, with one puberulent pistil, the stigmas deeply lobed, the bracts pale brown, scarious, pubescent, lobed; fruiting aments to 12 cm long, the capsules conic-elongated, acuminate, 2-valved, to 6 mm long, puberulent, on puberulent pedicels to 2 mm long, the dark brown seeds minute.

COMMON NAME: Large-toothed Aspen.

HABITAT: Disturbed areas in and around woodlands.

RANGE: Nova Scotia to North Dakota, south to Iowa, Tennessee, and North Carolina.

ILLINOIS DISTRIBUTION: Occasional to common in the northern half of the state, extending south to Marion and Wabash counties.

The large-toothed aspen is an attractive tree by virtue of its smooth, gray young trunks and its rustling leaves. The ability of the leaves to tremor at the slightest breeze is directly related to the flattened nature of the slender petiole. The light wood is a source of pulp in papermaking.

In autumn, the foliage turns a handsome pale yellow.

The large-toothed aspen flowers during April.

4. Populus × smithii Boivin, Nat. Canad. 93:435. 1966. *Fig. 31.*
Populus × barnesii Wagner, Mich. Bot. 9:54. 1970.

Trees to 20 m tall, with a trunk diameter up to 0.5 m, the crown round-topped, the bark smooth, gray, the twigs rather stout, brown,

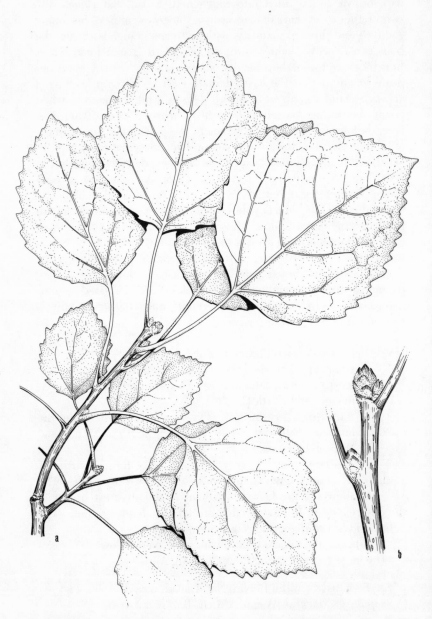

31. *Populus* × *smithii* (Hybrid Aspen). *a*. Vegetative branch, X¾. *b*. Winter twig, X1½.

pubescent at first, becoming glabrous at maturity, with scattered orange lenticels, the buds ovoid, acute, brown, pubescent, to 6 mm long; leaves broadly ovate to suborbicular, short-acuminate at the apex, more or less rounded at the base, dentate with 12–25 teeth (averaging 20), to 9 cm long, to 7.5 cm broad, pubescent when young, becoming glabrous or nearly so, dark green above, paler beneath, the petiole flat, up to 6 cm long, usually glabrous at maturity; stipules small, caducous; staminate aments pendulous, densely flowered, to 5 cm long, borne as the leaves begin to expand, the disk of each flower asymmetrical, entire, with 6–12 stamens, the filaments short, distinct, the bracts brown, pubescent, lobed; pistillate aments pendulous or spreading, sparsely flowered, to 8 cm long, borne as the leaves begin to expand, the disk of each flower asymmetrical, crenulate, with one puberulent pistil, the stigmas deeply lobed, the bracts brown, pubescent, lobed; fruiting aments to 10 cm long, the capsule conic-elongated, acuminate, 2-valved, to 6 mm long, puberulent, on puberulent pedicels to 2 mm long, the dark brown seeds minute.

COMMON NAME: Hybrid Aspen.

HABITAT: Woods.

RANGE: Michigan; northern Illinois.

ILLINOIS DISTRIBUTION: Known only from La Salle and Vermilion counties.

Populus × *smithii* is the hybrid between *P. grandidentata* Michx. and *P. tremuloides* Michx. It is intermediate in several characters, the most obvious being the number of dentations along the leaf margins. Wagner (1970) described this hybrid as *P.* × *barnesii*, failing to realize Boivin's earlier binomial for it.

This hybrid flowers during April.

5. Populus tremuloides Michx. Fl. Bor. Am. 2:243. 1803. *Fig. 32.*

Trees to 15 m tall, with a trunk diameter up to 0.4 m, the crown with a loose, rounded top, the bark pale yellow-green to white when young, becoming divided into black scaly ridges at maturity, the twigs slender, reddish-brown at first, becoming dark gray, nearly glabrous, with orange lenticels, the buds conic, acute, reddish-brown, nearly glabrous, to 10 mm long; leaves broadly ovate, short-acuminate at the apex, more or less rounded at the base, crenate-serrate with 20 or more glandular teeth (averaging 31), to 8 cm long,

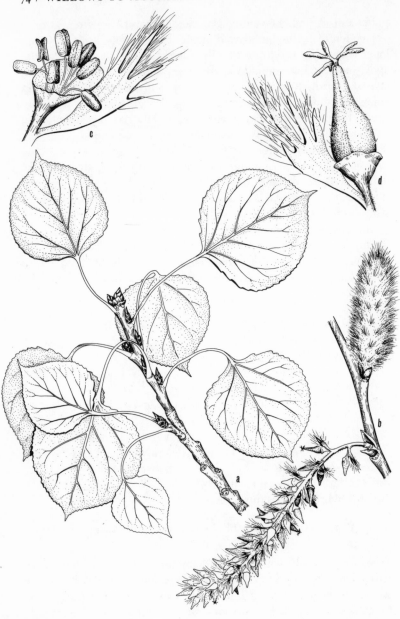

32. *Populus tremuloides* (Quaking Aspen). *a.* Vegetative branch, X½. *b.* Branch with staminate ament (above) and pistillate ament (below), X½. *c.* Staminate flower, X6. *d.* Pistillate flower, X6.

to 7 cm broad, glabrous on both surfaces, shiny green above, dull green beneath, the petiole flat, up to 6 cm long, glabrous; stipules small, caducous; staminate aments pendulous, densely flowered, to 6 cm long, borne as the leaves begin to expand, the disk of each flower asymmetrical, entire, with 6–12 stamens, the filaments short, distinct, the bracts brown, long-pubescent, deeply lobed; pistillate aments pendulous or spreading, sparsely flowered, to 8 cm long, borne as the flowers begin to expand, the disk of each flower asymmetrical, crenulate, with one puberulent pistil, the stigmas divided, the bracts brown, long-pubescent, deeply lobed; fruiting aments to 10 cm long, the capsule conic-elongated, acuminate, 2-valved, to 6 mm long, puberulent, on puberulent pedicels to 2 mm long, the light brown seeds minute.

COMMON NAME: Quaking Aspen.

HABITAT: Low ground of woods, marshes, and bogs.

RANGE: Newfoundland to Alaska, south to California, New Mexico, central Illinois, Tennessee, and New Jersey.

ILLINOIS DISTRIBUTION: Occasional in the northern half of the state, extending southward to Macoupin and Coles counties.

Quaking aspen has the ability to germinate readily in forests recently denuded by fires. It often forms large colonies because of its ability to produce suckers from the roots.

The wood is very light and soft, making it useful for pulp used in papermaking.

The flowers appear in late March and April.

6. Populus alba L. Sp. Pl. 1034. 1753. *Fig. 33.*

Trees to 20 m tall (in Illinois), with a trunk diameter up to 0.75 m, the crown open or broadly rounded, usually rather irregular, the bark greenish-gray to whitish at first, becoming darker and ridged with age, the twigs moderately stout, covered with dense, white, feltlike pubescence, at least while young; the buds ovoid, acute, canescent, to 8 mm long; leaves broadly ovate to suborbicular in outline, irregularly undulate-dentate or palmately 3- to 5-lobed, obtuse to acute at the apex, truncate to subcordate at the base, coriaceous, to 10 cm long, to 8 cm broad, densely white-tomentose on both surfaces when young, becoming more or less glabrous on the upper surface at maturity, dark green above, white beneath, the petiole terete, tomentose, to 3.5 cm long; stipules small, caducous;

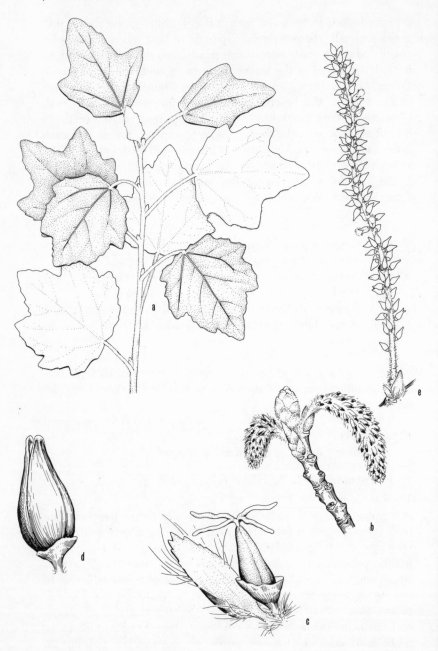

33. *Populus alba* (White Poplar). *a.* Vegetative branch, X½. *b.* Twig with aments, X1½. *c.* Pistillate flower, X7½. *d.* Fruiting ament, X⅔. *e.* Capsule, X7½.

staminate aments pendulous, densely flowered, to 10 cm long, borne as the leaves begin to expand, the disk of each flower more or less symmetrical, entire to undulate, with 6–12 stamens, the filaments short, distinct, the bracts brownish, pubescent with many short, triangular teeth; pistillate aments pendulous, less densely flowered, to 10 cm long, borne as the flowers begin to expand, the disk of each flower more or less symmetrical, usually crenulate, with one tomentose pistil, the bracts brownish, pubescent, with many short, triangular teeth; capsule oblongoid-ovoid, acuminate, 2-valved, to 5 mm long, tomentose, on very short, pubescent pedicels, the light brown seeds minute.

COMMON NAME: White Poplar; Abele.

HABITAT: Along roads; around abandoned homesites.

RANGE: Native of Europe; adventive in most parts of the eastern United States.

ILLINOIS DISTRIBUTION: Scattered throughout much of Illinois.

The white poplar, or abele, persists along roads and in fencerows. Several trees are usually found together, the result of the production of long, stoloniferous roots.

Much of the ornamental value of this species comes from the leaves which are dark green on the upper surface and velvety-white on the lower surface.

The flowers are borne from March to May.

7. **Populus canescens** (Ait.) Sm. Fl. Brit. 3:1080. 1805. *Fig. 34.*
Populus alba L. var. *canescens* Air. Hort. Kew. 3:405. 1789.

Trees to 15 m tall (in Illinois), with a trunk diameter up to 0.4 m, the crown more or less rounded, the trunk light gray when young, becoming dark ridged with maturity, the twigs moderately stout, pubescent, at least when young, the buds ovoid, acute, more or less canescent, to 6 mm long; leaves broadly ovate, undulate-dentate, never 3- to 5-lobed, usually acute at the apex, truncate at the base, to 8 cm long, to 6 cm broad, rather densely pubescent on both surfaces when young, becoming nearly glabrous on the upper surface and sparsely pubescent on the lower surface at maturity, green above, gray beneath, the petiole terete, sparsely pubescent, to 3 cm long; stipules small, caducous; staminate aments pendulous, densely flowered, to 8 cm long, borne as the leaves begin to expand, the disk of each flower more or less symmetrical, usually entire, with 6–12 stamens, the bracts brownish, pubescent, with 7–9 deep

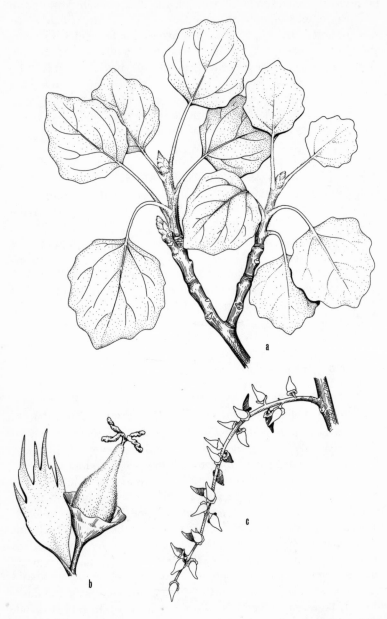

34. Populus canescens (Gray Poplar).   *a.* Vegetative branch, X½.   *b.* Pistillate flower, X6.   *c.* Fruiting ament, X1.

lacerations; pistillate aments pendulous, less densely flowered, to 8 cm long, borne as the leaves begin to expand, the disk of each flower more or less symmetrical, usually crenulate, with one tomentulose pistil, the bracts brownish, pubescent, with 7–9 lacerations; capsule oblongoid-ovoid, acuminate, 2-valved, to 5 mm long, tomentose, on very short, pubescent pedicels, the light brown seeds minute.

COMMON NAME: Gray Poplar.

HABITAT: Along roads.

RANGE: Native of Europe; infrequently spreading from cultivation in the eastern United States.

ILLINOIS DISTRIBUTION: Not common; known from Adams, Champaign, Lake, and Richland counties

Aiton described this plant in 1789 as a variety of *P. alba*. There are many botanists today who believe that this should still be the status accorded the gray poplar.

The basic differences exhibited by *P. canescens* are the lacerate bracts, the more sparsely pubescent leaves, and the absence of any lobing of the leaves.

The gray poplar is much less frequent in Illinois than the white poplar.

The flowers are borne in April and May.

8. Populus balsamifera L. Sp. Pl. 1034. 1753. *Fig. 35.*
Populus tacamahacca Mill. Gard. Dict. ed. 8, no. 6. 1768.

Trees to 15 m tall (in Illinois), with a trunk diameter up to 0.5 m, the crown mostly narrowly pyramidal, the trunk smooth, gray, at maturity shallowly fissured, the twigs stout, yellow-brown, usually glabrous, shiny, with bright orange lenticels, the buds ovoid, acute, to 2.5 cm long, extremely resinous with yellow balsam; leaves ovate-lanceolate, acute to acuminate at the apex, rounded to subcordate at the base, crenulate to serrulate, to 10 cm long, to 7 cm wide, at first strongly resinous, at length dark green above, pale green and more or less with a rusty metallic luster beneath, glabrous above, puberulent on the midvein beneath, the petiole terete, to 3 cm long, glabrous or nearly so; stipules small, caducous; staminate aments pendulous, densely flowered, to 8 cm long, borne as the leaves begin to expand, the disk of each flower symmetrical, entire, with 20–40 stamens, the bracts light brown, mostly glabrous, laciniate at the apex; pistillate aments pendulous, loosely flowered, to 10 cm long, borne as the leaves begin to expand, the disk of each

35. *Populus balsamifera* (Balsam Poplar). *a.* Branch with leaves and fruiting ament, X½. *b.* Branch with pistillate ament, X¼. *c.* Pistillate flower, X2½.

flower symmetrical, entire, with one glabrous pistil, the bracts light brown, mostly glabrous, laciniate at the tip; capsule oblongoid-ovoid, acute, 2-valved, to 8 mm long, glabrous, on very short, glabrous pedicels, the light brown seeds minute.

COMMON NAMES: Balsam Poplar; Tacamahac.

HABITAT: Dunes.

RANGE: Labrador to Alaska, south to Colorado, Nebraska, northeastern Illinois, and New York.

ILLINOIS DISTRIBUTION: Known only from Cook, Lake, and McHenry counties.

The balsam poplar is rare in Illinois, being restricted to dunal areas around Lake Michigan. The first report of it from Illinois was made apparently by Higley and Raddin in 1891. This is the same plant which sometimes has been known as *P. tacamahacca* Mill. These two binomials seemingly refer to the same species, however.

The leaves and buds are heavily saturated with a balsam resin which imparts a strong fragrance.

The similarly resinous *P.* × *gileadensis* has more pubescent, cordate leaves.

Populus balsamifera flowers during April.

9. **Populus** × **gileadensis** Rouleau, Rhodora 50:235. 1948. *Fig. 36.*

Populus candicans Ait. Hort. Kew. 3:406. 1789.

Populus balsamifera L. var. *candicans* (Ait.) Gray, Man. ed. 2, 419. 1856.

Tree to 20 m tall (shorter in Illinois), with a trunk diameter up to 0.3 m, the crown widely spreading, irregular, the younger bark yellow-brown, smooth, the older bark gray, with thick, firm ridges, the twigs stout, yellow-brown, more or less pubescent, shiny, with orange lenticels, the buds lanceoloid, acuminate, to 2.5 cm long, extremely resinous with yellow balsam; leaves broadly ovate, short-acuminate at the apex, cordate at the base, crenate to serrate, to 12 cm long, to 10 cm broad, at first pubescent, later becoming glabrous above, sparsely pubescent beneath, especially on the margins and the veins, dark green above, white and often rusty-metallic beneath, the petioles terete, usually sparsely pubescent, to 3 cm long; stipules small, caducous; staminate aments unknown; pistillate aments pendulous, loosely flowered, to 10 cm long, borne as the leaves begin to expand, the disk of each flower symmetrical, entire,

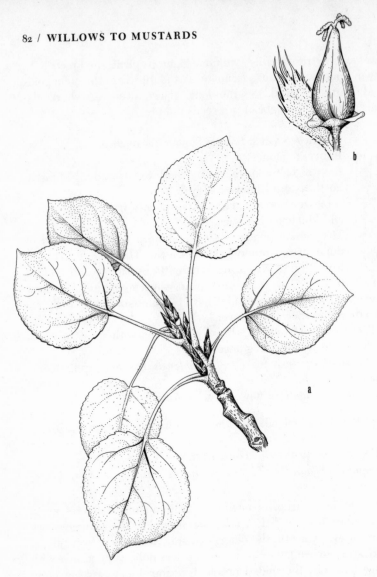

36. *Populus × gileadensis* (Balm-of-Gilead). *a.* Vegetative branch, X⅜. *b.* Pistillate flower, X6.

with one glabrous pistil, the bracts light brown, usually sparsely pubescent, laciniate at the tip; fruiting aments to 12 cm long, the capsules ovoid, acute, 2-valved, to 7 mm long, glabrous, on glabrous or sparsely pubescent pedicels to 3 mm long.

COMMON NAME: Balm-of-Gilead.

HABITAT: Roadsides.

RANGE: Native to Canada and perhaps the northernmost United States; escaped from cultivation southward.

ILLINOIS DISTRIBUTION: Known only from Lake County.

The origin of the balm-of-Gilead is not thoroughly understood. Since staminate aments have never been found, it is thought by some that it represents a hybrid between *P. balsamifera* and *P. deltoides*, a view followed in this work.

There are some botanists, however, who prefer to consider it a sterile clone of *P. balsamifera* L. var. *subcordata* Hylander. Still others designate it as a normal species, referring to it as *P. candicans* Ait.

Populus × *gileadensis* differs from the similar *P. balsamifera* by its cordate, more pubescent leaves.

The pistillate aments are borne in April.

10. Populus heterophylla L. Sp. Pl. 1034. 1753. *Fig. 37.*

Populus argentea Michx. f. Hist. Arb. Am. 3:290, pl. 9. 1813.

Trees to 25 m tall, with a trunk diameter up to 0.8 m, the crown more or less open, the trunk usually straight, columnar, the bark reddish-brown, broken into elongated plates, the twigs stout, gray or brown, somewhat shiny, tomentose when young, becoming sparsely pubescent or nearly glabrous with age, the buds ovoid, acute, to 1 cm long, reddish-brown, slightly resinous; leaves broadly ovate, obtuse to subacute at the apex, truncate, rounded, or subcordate at the base, glandular-serrate, to 20 cm long, to 15 cm broad, white-tomentose when young, becoming glabrous or nearly so at maturity, dark green above, paler below, the petiole terete, slender, yellow, more or less pubescent, to 8 cm long; stipules small, caducous; staminate aments erect at first, becoming pendulous, to 6 cm long, densely flowered, borne as the leaves begin to expand, the disk of each flower asymmetrical, entire, with 12–20 stamens, the bracts brown, glabrous or pubescent, fimbriate at the apex; pistillate aments pendulous, becoming erect by fruiting time, loosely flowered, to 5 cm long, borne as the leaves begin to expand, the disk of each flower symmetrical, long-toothed, with one glabrous pistil, the bracts brown, glabrous or pubescent, fimbriate at the upper end; fruiting aments to 6 cm long, more or less erect, the capsule ovoid, acute, 2- to 3-valved, to 12 mm long, on glabrous

37. *Populus heterophylla* (Swamp Cottonwood). *a.* Branch with leaves and fruiting ament, X½. *b.* Twig with staminate aments, X1. *c.* Staminate flower, X2½. *d.* Pistillate flower, X2½.

pedicels usually longer than the capsules, with numerous minute seeds.

COMMON NAME: Swamp Cottonwood.

HABITAT: Swampy woods.

RANGE: Connecticut, along the Atlantic coast to northern Florida, across to eastern Texas, up the Mississippi River to southern Missouri, southern Illinois, and southern Indiana.

ILLINOIS DISTRIBUTION: Confined to the southern one-third of the state.

Swamp cottonwood is a common member of the bald cypress swamp community in extreme southern Illinois.

The lobed disk of the pistillate flowers is one of its most unique characters, as are its more or less erect fruiting aments.

The wood of the swamp cottonwood is used for pulp in paper-making and for the manufacture of excelsior.

The flowers are borne in March and April.

Order *Tamaricales*

The Tamaricales occupies a position in the Thorne (1968) classification between the Salicales and the Capparidales. It consists only of the family Tamaricaceae.

TAMARICACEAE – TAMARISK FAMILY

Shrubs or trees or rarely herbs; leaves alternate, simple, often scalelike, without stipules; flowers bisexual or rarely unisexual, actinomorphic, usually borne in spikes, panicles, or racemes, or sometimes solitary; sepals 4–5, free; petals 4–5, free; stamens 4–5, 8, or 10, attached to or below a disk; ovary superior, 1-locular, with basal or parietal placentation; fruit a capsule with many seeds.

Four genera comprise this family, all of them native to the Old World.

Only the following genus occurs in Illinois.

1. *Tamarix* L. – Tamarisk

Shrubs or small trees; leaves alternate, often scalelike, without stipules; inflorescence racemose or paniculate, with numerous small flowers; sepals 4–5, free; petals 4–5, free; stamens 4–5; disk present; ovary superior, with basal placentation; fruit a capsule, separating at maturity into 3–4 valves.

Of the seventy-five species of *Tamarix* in Europe and Asia which make up this genus, only the following has been found as an adventive in Illinois.

1. **Tamarix gallica** L. Sp. Pl. 270. 1753. *Fig. 38.*

Shrub or small tree; leaves alternate, minute, scalelike, pale green or even glaucous; inflorescence a terminal spike with numerous small flowers; sepals 5, green, free; petals 5, pink, free; stamens 5, each one arising from a sinus of the 5-lobed disk; capsule conical, containing numerous seeds.

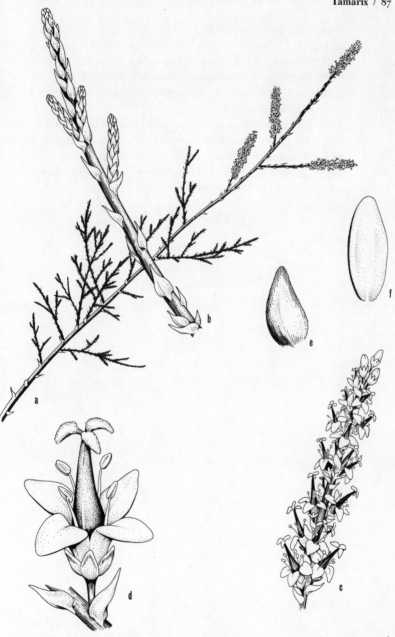

38. *Tamarix gallica* (Tamarisk). *a.* Branch with leaves and inflorescences, X½.
b. Branch with leaves and immature inflorescences, X3. *c.* Inflorescence, X3½.
d. Flower, X10. *e.* Sepal, X20. *f.* Petal, X20.

COMMON NAME: Tamarisk.

HABITAT: Disturbed ground, usually near water.

RANGE: Native of Europe; naturalized frequently in the western states, extending eastward at least to Illinois.

ILLINOIS DISTRIBUTION: Known from Jackson, Macoupin, and St. Clair counties.

This species, unusual in appearance because of its numerous, pale green, scalelike leaves, is occasionally grown as an ornamental in Illinois. It has apparently become established as an adventive in at least three localities in Illinois.

Tamarisk flowers from June to September.

Order Capparidales

The Capparidales forms the fourth and last order of the superorder Cistiflorae, according to the Thorne (1968) system of classification. It encompasses the families Capparidaceae, Moringaceae, Resedaceae, and Brassicaceae (= Cruciferae). All but the Moringaceae are represented in the Illinois flora.

Basically the order is held together on the basis of its polypetaly, its parietal placentation, and its usually 2-carpellate condition. More recently, it has been found that most members of the order possess specialized myrosin cells which produce the enzyme myrosin, an ingredient in mustard oil.

Although the spelling of the order has been conserved in the Code of Nomenclature as Capparales, I am in agreement with Crosswhite and Iltis's (1966) reasoning that the spelling should be Capparidales.

The Cronquist (1968) classification also groups the four families recognized here in his Capparales, but adds a fifth family, the Tovariaceae, which he splits out from the Capparidaceae.

Under the old Englerian system, these families, together with the Papaveraceae, comprised the order Rhoeadales.

CAPPARIDACEAE –CAPER FAMILY

Herbs (in Illinois), shrubs, trees, or woody vines; leaves alternate, simple or palmately compound, usually with minute stipules; inflorescence in bracteate racemes, the flowers mostly bisexual, actinomorphic or commonly zygomorphic; sepals 4–8, free or connate; petals 4–several or absent, free; stamens 4–several, free; pistil 1, usually borne on an elongated gynophore, the ovary superior, 1-locular, mostly 2-carpellate, the placentation parietal, with few to many ovules; fruit a capsule (in Illinois), berry, or nut.

With the inclusion of the tropical American genus *Tovaria*, the Capparidaceae is composed of 47 genera and about 800 species. They are found in the warmer regions of both the eastern and western hemispheres.

The presence of a gynophore is distinctive, as is the usually zygomorphic flower with 6 (in Illinois) or more stamens.

Most botanists are in agreement that the Capparidaceae and Brassicaceae are fairly close relatives.

The genus *Cleome* has a number of garden ornamentals, including the common giant spiderflower. Dried flower buds of *Capparis spinosa* L., the caper-bush of the Mediterranean region, are the source of the caper used in food seasoning.

In following Iltis (1958) on the alignment of genera, I am including *Cristatella* in *Polanisia*.

KEY TO THE GENERA OF Capparidaceae IN ILLINOIS

1. Stamens (6–) 7–16 (–32) _____ 1. *Polanisia*
1. Stamens 6 _____ 2. *Cleome*

1. *Polanisia* Raf. – Clammyweed

Annual or perennial herbs, often viscid-pubescent, with a rank odor; leaves trifoliolate; stipules minute; inflorescence a terminal, bracteate raceme; sepals 4, free; petals 4, clawed, two of them longer than the others; stamens (6–) 7–16 (–32), free, unequal in length, with a prominent gland between the stamens and the petals; pistil one, the ovary superior, the style slender; fruit an elongated capsule, dehiscing by valves from the apex, with several nearly spherical seeds.

The genus is composed of four species, according to Iltis (1958). Included are the species which generally have been placed in *Cristatella* Nutt.

KEY TO THE SPECIES OF Polanisia IN ILLINOIS

1. Leaflets obovate to lance-elliptic, all or most of them over 6 mm broad; fruits 5–10 mm broad _____ 1. *P. dodecandra*
1. Leaflets linear, up to 4 mm broad; fruits 3–4 mm broad _____
_____ 2. *P. jamesii*

1. Polanisia dodecandra (L.) DC. Prodr. 1:242. 1824.

Cleome dodecandra L. Sp. Pl. 2:672. 1753.

Annual herbs from a slender taproot; stem erect, to 60 cm tall, glancular-viscid; leaves trifoliolate, alternate, the petioles to 6 cm long, glandular-viscid; leaflets ovate or obovate to lance-elliptic, obtuse to acute at the apex, cuneate at the base, to 6 cm long, to 2 cm broad, entire, glandular-viscid; racemes terminal, many-flowered, the bracts unifoliolate or trifoliolate; sepals 4, free, lanceolate to ovate, green, to 1 cm long; petals 4, free, spatulate to obcordate, white or pink, to 1.6 cm long; stamens 11–16, free, purple, to 30 mm long; nectary (between stamens and petals) 1–2 mm tall, orange-tipped with yellow sides; pistil one, the ovary superior, the style to 40 mm long; gynophore to 9 mm long; capsule oblong to linear-fusiform, inflated, glandular-viscid, to 10 cm long, 5–10 mm broad, the valves reticulate, separating for ⅓–⅔ their length, the stipe 15–45 mm long; seeds several, nearly spherical, 1.7–3.1 mm in diameter, usually wrinkled.

Ernst (1963) and Iltis (1954, 1966) have thoroughly discussed the proper binomial for this species, Ernst maintaining it should be *P. graveolens* Raf. Iltis's overwhelming evidence in favor of *P. dodecandra* (L.) DC. is adhered to in this work.

Two subspecies are recognized, separated by the following key:

1. Largest petals 3.5–6.5 (–8.0) mm long; longest stamens 4–10 (–14) mm
    long, scarcely exceeding the petals _____
    _____1a. *P. dodecandra* ssp. *dodecandra*
1. Largest petals (7–) 8–16 mm long; longest stamens (9–) 12–30 mm
    long, usually much exceeding the petals _____
    _____1b. *P. dodecandra* ssp. *trachysperma*

1a. **Polanisia dodecandra** (L.) DC. ssp. **dodecandra** *Fig.* 39.
Polanisia graveolens Raf. Am. Jour. Sci. 1:379. 1819.

Largest petals 3.5–6.5 (–8.0) mm long; longest stamens 4–10 (–14) mm long, scarcely exceeding the petals.

COMMON NAME: Clammyweed.
HABITAT: Mostly along railroads.
RANGE: Quebec to Manitoba, south to Nebraska, Tennessee, and Maryland.
ILLINOIS DISTRIBUTION: Scattered throughout much of the state.

Iltis (1966) has placed what generally had been known as *P. graveolens* Raf. and *P. trachysperma* Torr. & Gray as subspecies of *P. dodecandra*. Thus *P. graveolens* becomes the typical subspecies of *P. dodecandra*.

The characters which distinguish the two subspecies, as indicated in the key, are not always clearly reliable.

This subspecies is almost always found in the ballast along railroads. It flowers from June to September.

1b. **Polanisia dodecandra** (L.) DC. ssp. **trachysperma** (Torr. & Gray) Iltis, Rhodora 68:47. 1966.
Polanisia trachysperma Torr. & Gray, Fl. N. Am. 1:669. 1840.
Polanisia dodecandra (L.) DC. var. *trachysperma* (Torr. & Gray) Iltis, Brittonia 10:44. 1958.

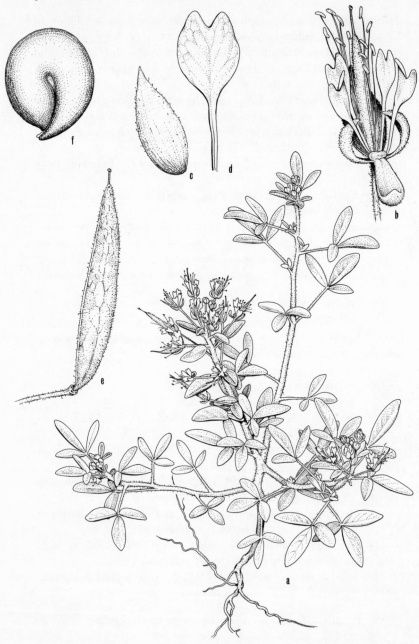

39. *Polanisia dodecandra* (Clammyweed). *a.* Habit, X¼. *b.* Flower, X1½. *c.* Sepal, X3. *d.* Petal, X3. *e.* Fruit, X¾. *f.* Seed, X10.

Largest petals (7–) 8–16 mm long; longest stamens (9–) 12–30 mm long, usually much exceeding the petals.

COMMON NAME: Clammyweed.
HABITAT: Sandy soil; gravelly soil.
RANGE: Saskatchewan to British Columbia, south to California, Texas, and Illinois.
ILLINOIS DISTRIBUTION: Scattered in all parts of the state, but not common.
This subspecies is much less common in Illinois than ssp. *dodecandra.*
Both subspecies give off a very unpleasant odor.
Polanisia dodecandra ssp. *trachysperma* flowers from June to September.

2. **Polanisia jamesii** (Torr. & Gray) Iltis, Brittonia 10:54. 1958.
Fig. 40.
Cristatella jamesii Torr. & Gray, Fl. N. Am. 1:124. 1838.

Annual herbs from a slender taproot; stem erect, to 40 cm tall, densely glandular-viscid; leaves trifoliolate, alternate, the petioles to 1.5 cm long, glandular-viscid; leaflets linear to linear-elliptic, conduplicate, acute and mucronulate at the apex, cuneate at the base, to 4 cm long, to 4 mm broad, entire, glandular-viscid; racemes terminal, relatively few-flowered, the bracts trifoliolate; sepals 4, united at the base, lanceolate, green, to 3 mm long; petals 4, free, spatulate, white or yellowish-white, unequal, the two larger ones 4–5 mm long and shallowly laciniate, the two smaller ones 2–3 mm long and deeply laciniate; stamens 6–9, pink, to 6 mm long; nectary tubular, yellow when fresh, 2.0–3.5 mm long; pistil one, the ovary superior, the style to 2 mm long, persistent on the fruit; gynophore 2.0–4.5 mm long; capsule linear-fusiform, slightly inflated, to 3 cm long, 3–4 mm broad, glandular-viscid, the valves separating nearly to the base, the stipe 8–10 mm long; seeds several, nearly spherical, 1.7–1.8 mm in diameter, reddish-brown, minutely pebbled.

COMMON NAME: Small Clammyweed.
HABITAT: Sandy soil along rivers; sandy blowouts.
RANGE: Northwestern Wisconsin to central Colorado, south to west-central Texas and central Illinois.
ILLINOIS DISTRIBUTION: Carroll, Jo Daviess, Mason, Tazewell, and Whiteside counties.
Before Iltis's study in 1958, this species had been segregated into its own genus, being referred to as *Cristatella*

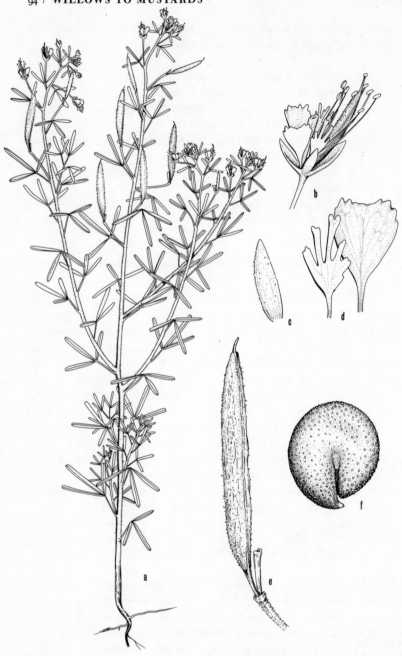

40. *Polanisia jamesii* (Small Clammyweed). *a.* Habit, X⅓. *b.* Flower, X2½. *c.* Sepal, X10. *d.* Petals, X7½. *e.* Fruit, X2½. *f.* Seed, X15.

jamesii Torr. & Gray, this on the basis of its few (6–9) stamens and laciniate petals.

This is an exceedingly rare species, confined to sandy shores, sandy blowouts in prairies, and sand dunes and sand hills. The first Illinois collection was made by H. A. Gleason and F. C. Gates in sandy blowouts near Hanover, Jo Daviess County, on June 17, 1908.

The small clammyweed flowers from mid-June to late September.

2. *Cleome* L. – Spiderflower

Mostly herbaceous annuals or perennials, sometimes viscid-pubescent, with a rank odor; leaves simple or more commonly 3- to 7-foliolate; stipules minute; inflorescence a bracteate raceme; sepals 4, free or united, sometimes persistent; petals 4, clawed, usually about equal in size; stamens 6, free, nearly equal, inserted on the receptacle above the petals, with a gland between the stamens and the pistil; pistil one, the ovary superior, stalked, the style slender; fruit an elongated capsule, stipitate, with many seeds.

There are about 75 species in this genus, mostly native to the warmer regions of the world.

KEY TO THE SPECIES OF Cleome IN ILLINOIS

1. Plants glabrous; leaflets 3; petioles without spines at the base _____ _____1. *C. serrulata*
1. Plants viscid-pubescent; leaflets 5 or 7; petioles with short spines at the base _____2. *C. hassleriana*

1. **Cleome serrulata** Pursh, Fl. Am. Sept. 441. 1814. *Fig. 41.*
Cleome integrifolia Torr. & Gray, Fl. N. Am. 1:122. 1838.

Annual herbs from a slender root; stem erect, to about 1 m tall, branching at least above, glabrous; leaves trifoliolate, alternate, the petioles of the lower leaves long, those of the uppermost leaves absent, glabrous; leaflets elliptic to lanceolate, acute at the apex, cuneate at the base, to 5 cm long, about ⅓ as broad, entire or occasionally serrulate, glabrous; racemes terminal, several-flowered, the bracts unifoliolate, linear to lanceolate; flowers pink or white, on slender pedicels; sepals 4, free, green; petals 4, free, oblong, obtuse, white or pink, up to 1.5 cm long, with a short claw; stamens 6, long-exserted beyond the petals; pistil one, the ovary superior, borne on a gynophore; capsule linear, pointed at the tip, glabrous, to 5 cm long, to 5 mm broad.

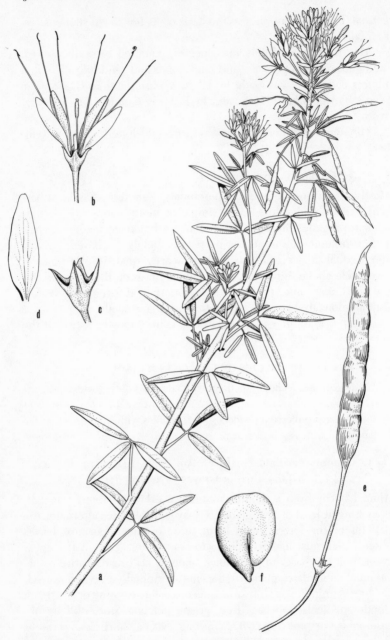

41. *Cleome serrulata* (Stinking Clover). *a.* Upper part of plant, X½. *b.* Flower, X1½. *c.* Calyx, X2. *d.* Petal, X2. *e.* Fruit, X1½. *f.* Seed, X10.

COMMON NAME: Stinking Clover; Pink Cleome.

HABITAT: Adventive in Illinois in waste areas.

RANGE: Manitoba to Washington, south to Arizona and Missouri; adventive in Illinois.

ILLINOIS DISTRIBUTION: Scattered throughout the state. Although Fernald (1950) and others indicate Illinois to be in the natural range of this species, most if not all the Illinois collections have been from waste areas.

Illinois botanists before the twentieth century called this species *C. integrifolia*, but Pursh's binomial clearly predates this.

Pink cleome is different from the other cleome in Illinois by its trifoliolate leaves and absence of short spines at the base of the petioles.

This species blooms from late May to September.

2. Cleome hassleriana Chod. Bull. Herb. Bioss. 6:App. 1, 12. *Fig. 42.*

Annual herbs from a slender root; stem erect, to 1.5 m tall, branching at least above, viscid-pubescent; leaves 5- to 7-foliolate, alternate, the lower on longer petioles than the upper, the petioles spiny at the base; leaflets lanceolate, acute at the apex, cuneate at the base, to 6 cm long, to 2 cm broad, serrulate, viscid-pubescent; racemes terminal, several-flowered, the bracts unifoliolate, lanceolate to ovate, cordate, sessile or nearly so; flowers pink or white, on long pedicels; sepals 4, free, green; petals 4, free, obovate, obtuse, up to 2 cm long, with a long claw; stamens 6, long-exserted beyond the petals; pistil one, the ovary superior, borne on a gynophore; capsule linear, acute or obtuse at the tip, glabrous, to 12 cm long, to 1 cm broad, the stipe up to 15 cm long.

COMMON NAME: Spiderflower.

HABITAT: Waste ground.

RANGE: Native to the Old World; adventive in eastern North America.

ILLINOIS DISTRIBUTION: Scattered throughout the state. Specimens which have been reported previously in Illinois as *Cleome spinosa* L. and *Cleome speciosissima* Deppe have been determined by Iltis to be *Cleome hassleriana*. Therefore, I am excluding the former two species from Illinois.

Spiderflower, which is a frequently cultivated species in Illinois

42. Cleome hassleriana (Spiderflower). *a.* Upper part of plant, X⅓. *b.* Flower, X1¼. *c.* Sepal, X2½. *d.* Petal, X2½. *e.* Fruit, X½. *f.* Seed, X12½. *g.* Flower with blade of petals removed to show attachment of floral parts to receptacle, X2.

gardens, has been found as an escape in waste areas in several parts of the state.

This species flowers from July to September.

RESEDACEAE – MIGNONETTE FAMILY

Herbs, rarely becoming woody near the base; leaves alternate, simple to pinnately compound, with glandular stipules; inflorescence racemose or spicate, with several bracteate flowers, the flowers asymmetrical and bisexual; sepals (4–) 5–6 (–7), united below; petals (4–) 5–6 (–7), united below; disk fleshy, asymmetrical; stamens 3–40, attached to the disk; ovary one, superior, 3- to 6-locular, the placentation parietal; fruit a lobed capsule, opening at the top.

Of the six genera which make up this family, only the genus *Reseda* has been found in Illinois.

1. Reseda [*Tourn.*] L. – Mignonette

Herbs; leaves alternate, simple to pinnately compound; inflorescence racemose or spicate, with several small bracteate flowers; sepals 4–7, united below; petals 4–7, united below; stamens 10–40, attached to one side of the flower; ovary superior; fruit a lobed capsule, opening at the top.

There are about 60 Old World species in this genus, with only the following known from Illinois.

1. Reseda alba L. Sp. Pl. 449. 1753. *Fig. 43*.

Herbs; stems erect, glabrous, somewhat glaucous, up to 1 m tall; leaves alternate, pinnately compound to pinnatifid, with 9–13 leaflets or lobes, the leaflets or lobes lanceolate to oblong to linear, obtuse at the apex, cuneate at the base, entire, glabrous; inflorescence a dense, spikelike raceme; flowers white, to 5 mm across, on short pedicels; sepals 6-parted, green; petals 6-parted, each lobe 3-cleft at the tip; stamens 12–15; ovary superior; capsule ovoid to oblongoid, 4-toothed at the apex, to 1 cm long.

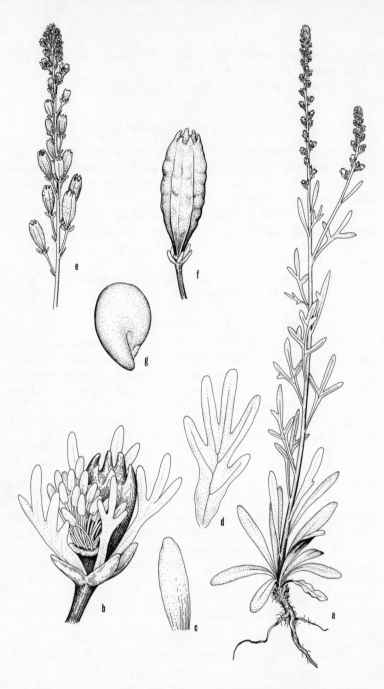

43. *Reseda alba* (Mignonette). *a.* Habit, X¼. *b.* Flower, X10. *c.* Sepal, X12½.
d. Petal, X12½. *e.* Fruiting raceme, X1½. *f.* Fruit, X5. *g.* Seed, X12½.

COMMON NAME: Mignonette.

HABITAT: Disturbed areas.

RANGE: Native of Europe; adventive in North America.

ILLINOIS DISTRIBUTION: Known only from Cook and Du Page counties.

Mignonette was a favorite plant in flower gardens in years past, but is not often seen today. Two records of this species as an adventive are known from Cook and Du Page counties, one of them collected as early as 1898.

The flowers bloom in July and August.

BRASSICACEAE – MUSTARD FAMILY

Herbs, often with stellate pubescence; leaves alternate, simple or pinnatifid or palmately lobed, without stipules; inflorescence mostly in bractless racemes, the flowers bisexual, usually actinomorphic; sepals 4, free, in 2 whorls; petals 4, free; stamens mostly 6, free, the outer whorl of two usually shorter than the inner whorl of four; pistil one, the ovary superior, 2-locular and 2-carpellate, the 2 locules separated by a membranous replum, the placentation parietal, with few to several ovules; fruit dehiscent, valvate, either an elongated silique or a short silicle; seeds several.

The Brassicaceae is composed of about 350 genera and 2500 species, mostly in the cooler regions of the northern hemisphere.

Since I am adhering to the suffix -aceae for all families, I am using the family name Brassicaceae in place of the more familiar Cruciferae.

The Brassicaceae is one of the most easily recognized families because of the four free petals, six stamens of two different lengths, and a fruit separated into two compartments by a membranous partition known as a *replum*. In addition, the leaves are invariably alternate.

Although many members of the family are among our most widely distributed weeds, a number of genera are grown for the handsome flowers. Included among the garden ornamentals are the stocks (*Matthiola*), candytuft (*Iberis*), alyssum (*Alyssum*), sweet alyssum (*Lobularia*), rocket (*Hesperis*), and wallflowers (*Cheiranthes* and *Erysimum*). The moonwort (*Lunaria*) is grown for its fruit which resembles silver dollars, due to the exposed shiny replum evident after the valves drop off.

More important, however, are those genera grown for vegetables. This includes *Brassica*, the source of cabbage, broccoli, cauliflower, turnips, Brussels sprouts, kohlrabi, and rutabaga, *Raphanus* (radish), and *Nasturtium* (water cress). In addition, *Armoracia* is the source of horseradish.

KEY TO THE GENERA OF Brassicaceae IN ILLINOIS

1. Flowers white or purple _____ 2
1. Flowers yellow _____ 43
    2. Leaves deeply palmately divided _____ 1. *Dentaria*
    2. Leaves simple, pinnately divided, or pinnatifid _____ 3
3. Flowers purplish or reddish _____ 4
3. Flowers white _____ 12
    4. Plants pubescent _____ 5
    4. Plants glabrous _____ 9
5. Some or all the leaves pinnatifid _____ 6
5. None of the leaves pinnatifid _____ 7
    6. Plants hispid; roots thickened; petals 1.5 cm long or longer; silique 5–10 mm in diameter _____ 2. *Raphanus*
    6. Plants more or less glandular-hirtellous; roots not conspicuously thickened; petals less than 1.5 cm long; silique up to 2 mm in diameter _____35. *Chorispora*
7. None of the leaves cordate; fruits at least 5 cm long, somewhat constricted between the seeds _____ 3. *Hesperis*
7. At least the lowermost leaves cordate; fruits usually at most up to 4 cm long, not constricted between the seeds _____ 8
    8. Fruits narrowly lanceoloid, to 3 mm broad; seeds oval; lowermost leaves up to 2 cm across _____ 10. *Cardamine*
    8. Fruits broadly oval, about 2 cm broad; seeds suborbicular; lowermost leaves more than 2 cm across _____ 38. *Lunaria*
9. At least some of the upper leaves auriculate at the base _____
    _____4. *Iodanthus*
9. None of the leaves auriculate at the base _____ 10
    10. Plants fleshy; leaves sinuate-dentate; flowers 4–6 mm across _____
    _____ 5. *Cakile*
    10. Plants not fleshy; leaves entire or some or all of the leaves pinnatifid; flowers 10 mm or more across _____ 11
11. Some or all of the leaves pinnatifid; plants green _____ 6. *Eruca*
11. Leaves never pinnatifid, usually entire; plants gray or glaucous _____
    _____37. *Matthiola*
    12. Basal rosette present (although sometimes withered at flowering time) _____13
    12. Basal rosette absent _____ 24
13. Some of the leaves pinnate or pinnatifid _____ 14

13. None of the leaves pinnate or pinnatifid _____ 21
    14. Some or all the cauline leaves sagittate or auriculate at the base _____ 15
    14. None of the cauline leaves sagittate or auriculate at the base ____ _____ 16
15. Ovary and fruit triangular, broader than long _____ 7. *Capsella*
15. Ovary and fruit not triangular, much longer than broad _____ _____ 8. *Arabis*
    16. Leaves bi- or tri-pinnate; pubescence stellate _____ _____ 9. *Descurainia*
    16. Leaves once-pinnate or merely pinnatifid; pubescence simple (hairs bifurcate in *A. lyrata*) _____ 17
17. Petals 5 mm long or longer _____ 18
17. Petals up to 5 mm long _____ 20
    18. Siliques falcate, usually 5–10 cm long; seeds winged _____ _____ 8. *Arabis*
    18. Siliques straight, up to 4.5 cm long; seeds wingless _____ 19
19. Flowers 6–9 mm broad; petals 5–8 mm long; leaves hirsute, the uppermost simple _____ 8. *Arabis*
19. Flowers 10–15 mm broad; petals 8–15 mm long; leaves glabrous, all of them pinnate _____ 10. *Cardamine*
    20. Silique beaked; seeds wingless _____ 10. *Cardamine*
    20. Silique nearly beakless; seeds winged _____ 11. *Sibara*
21. Ovary and fruit only as broad as long or at most 2–7 times longer than broad _____ 22
21. Ovary and fruit much greater than seven times as long as broad _____ _____ 23
    22. Ovary and fruit longer than broad; plants pubescent _____ _____ 12. *Draba*
    22. Ovary and fruit as broad as long; plants glabrous _____ _____ 17. *Thlaspi*
23. Leaves of basal rosette never exceeding 1 cm in width _____ _____ 13. *Arabidopsis*
23. Most leaves of basal rosette 1 cm broad or broader _____ 8. *Arabis*
    24. Some or all the leaves pinnate or pinnatifid _____ 25
    24. None of the leaves pinnate or pinnatifid _____ 31
25. Some or all the cauline leaves auriculate or sagittate _____ 26
25. None of the cauline leaves auriculate or sagittate _____ 27
    26. Plants usually pubescent; fruits with 1 seed per cell _____ _____ 14. *Lepidium*
    26. Plants glabrous; fruits with 2–8 seeds per cell _____ 17. *Thlaspi*
27. Petals to 2 mm long, or absent _____ 28

27. Petals 2–20 mm long _____ 29
    28. Plants erect; silicle flattened _____ 14. *Lepidium*
    28. Plants matted; silicle subglobose _____ 36. *Coronopus*
29. Petals 2–8 mm long; plants glabrous _____ 30
29. Petals 10–20 mm long; plants more or less hispid _____ 2. *Raphanus*
    30. Some or all the cauline leaves simple; fruits up to three times as long as broad _____ 15. *Armoracia*
    30. All cauline leaves pinnately compound; fruits more than three times as long as broad _____ 16. *Nasturtium*
31. Some or all the cauline leaves clasping or sagittate _____ 32
31. None of the leaves clasping or sagittate _____ 38
    32. Petals 7–20 mm long _____ 33
    32. Petals up to 6 mm long _____ 34
33. Basal leaves on petioles longer than the blades; siliques 1.5–3.5 cm long _____10. *Cardamine*
33. Basal leaves on petioles shorter than the blades; siliques 4–10 cm long _____8. *Arabis*
    34. Cauline leaves glabrous _____ 35
    34. Cauline leaves pubescent _____ 36
35. Ovary and fruit nearly orbicular, the fruit 1.0–1.8 cm in diameter _____ 17. *Thlaspi*
35. Ovary and fruit long and slender, the fruit 5–10 cm long _____ _____8. *Arabis*
    36. Ovary and fruit not more than three times longer than broad ____ _____37
    36. Ovary and fruit many times longer than broad _____ 8. *Arabis*
37. Fruit winged, notched at apex, 2–3 times longer than broad; petals up to 2 mm long _____ 14. *Lepidium*
37. Fruit unwinged, not notched at apex, as broad as long; petals 3–4 mm long _____18. *Cardaria*
    38. Petals 15–20 mm long _____ 15. *Armoracia*
    38. Petals less than 15 mm long _____ 39
39. Leaves broadly ovate, the lowermost on petioles nearly as long as or longer than the blades; plants with the odor of garlic _____ _____19. *Alliaria*
39. Leaves variously shaped, but not broadly ovate, the lowermost sessile or on short petioles; plants without the odor of garlic _____ 40
    40. Cauline leaves entire _____ 41
    40. Cauline leaves toothed _____ 42
41. Petals deeply notched; pubescence of stellate hairs _____ _____20. *Berteroa*
41. Petals entire; pubescence of bifurcate hairs _____ 21. *Lobularia*

42. Ovary and fruit about as broad as long; petals up to 2 mm long; stems to 45 cm tall _____ 14. *Lepidium*

42. Ovary and fruit much longer than broad; petals 4–6 mm long; stems 1–2 m tall _____ 8. *Arabis*

43. Basal rosette present (sometimes rather withered at flowering time) _____ 44

43. Basal rosette absent _____ 49

44. Some or all the cauline leaves auriculate or sagittate at the base _____ 45

44. None of the cauline leaves auriculate nor sagittate at the base _____ 47

45. Basal leaves deeply pinnatifid; cauline leaves pinnatifid or coarsely dentate; petals yellow _____ 46

45. Basal leaves lyrate or merely toothed; cauline leaves mostly entire; petals creamy-yellow _____ 8. *Arabis*

46. Petals 5 mm long or longer; ovaries and fruits several times longer than broad _____ 22. *Barbarea*

46. Petals up to 2 mm long; ovaries and fruits at most about three times longer than broad _____ 34. *Rorippa*

47. Basal leaves linear to elliptic, not pinnatifid or pinnate _____ _____ 23. *Lesquerella*

47. Basal leaves pinnatifid or pinnate _____ 48

48. Basal leaves pinnatifid; petals 6–10 mm long _____ 24. *Diplotaxis*

48. Basal leaves bi- or tri-pinnate; petals 2.0–2.5 mm long _____ _____ 9. *Descurainia*

49. None of the leaves pinnatifid nor pinnate _____ 50

49. Some or all the leaves pinnate or pinnatifid _____ 58

50. None of the leaves auriculate nor sagittate at the base _____ 51

50. Some of the leaves auriculate or sagittate at the base _____ 53

51. Ovary and fruit many times longer than broad; pubescence of 2- to 3-parted hairs _____ 25. *Erysimum*

51. Ovary and fruit about as broad as long; pubescence of stellate hairs _____ 52

52. Petals up to 4 mm long; silicle 4-seeded; leaves up to 3 cm long _____ 26. *Alyssum*

52. Petals 4–7 mm long; silicle several-seeded; most of the leaves over 3 cm long _____ 23. *Lesquerella*

53. Plants glabrous _____ 54

53. Plants pubescent _____ 57

54. Style absent; fruits borne on pendulous pedicels _____ 27. *Isatis*

54. Style elongate; fruits borne on ascending pedicels _____ 55

55. None of the leaves on the stem entire _____ 22. *Barbarea*

55. Some or all of the leaves on the stem entire _____ 56
    56. Petals 7–10 mm long; cauline leaves obtuse at apex _____
    _____28. *Conringia*
    56. Petals up to 6 mm long; cauline leaves acute at apex _____
    _____ 29. *Camelina*
57. Petals 2–3 mm long; fruit 1-seeded _____ 30. *Neslia*
57. Petals 4–6 mm long; fruit several-seeded _____ 29. *Camelina*
    58. Some of the leaves auriculate at the base _____ 59
    58. None of the leaves auriculate at the base _____ 61
59. Ovary and fruit several times longer than broad _____ 60
59. Ovary and fruit at most only about three times longer than broad ____
    _____ 34. *Rorippa*
    60. Fruit tapering to a slender beak; seeds globose _____ 31. *Brassica*
    60. Fruit tapering to a short, thick style; seeds not globose _____
    _____ 22. *Barbarea*
61. Plants with stellate pubescence _____ 9. *Descurainia*
61. Plants glabrous or with pubescence of simple hairs _____ 62
    62. Racemes subtended by pinnatifid bracts _____ 32. *Erucastrum*
    62. Racemes without pinnatifid bracts subtending them (except rarely
    the lowest raceme) _____ 63
63. Petals up to 1 cm long _____ 64
63. Petals 1–2 cm long _____ 67
    64. Ovary and fruit conspicuously beaked _____ 65
    64. Ovary and fruit beakless _____ 66
65. Seeds in two rows _____ 24. *Diplotaxis*
65. Seeds in one row _____ 31. *Brassica*
    66. Ovary and fruit at least 15 times longer than broad; seeds in one
    row _____33. *Sisymbrium*
    66. Ovary and fruit up to 7 times longer than broad; seeds in two rows
    _____ 34. *Rorippa*
67. Petals with brown or purple veins _____ 68
67. Petals without brown or purple veins _____ 31. *Brassica*
    68. Pedicels up to 6 mm long; fruit not constricted between the seeds
    _____ 6. *Eruca*
    68. Pedicels over 6 mm long; fruit more or less constricted between
    the seeds _____ 2. *Raphanus*

1. *Dentaria* L.–Toothwort

Herbs with fleshy rhizomes; stem leafless except for a whorl of palmately divided leaves at or above the middle; basal leaves usually
several, palmately divided; inflorescence in corymbs or short racemes, the flowers actinomorphic; sepals 4, free; petals 4, free,

much longer than the sepals; stamens 6, free; pistil 1, the ovary superior, the style slender; fruit a silique, flat, dehiscent from the base, with wingless seeds in one row in each cell of the ovary.

Dentaria is a genus of about fifteen woodland herbs, all native to the northern hemisphere.

Only the following species occurs in Illinois.

1. **Dentaria laciniata** Muhl. ex Willd. Sp. Pl. 3:499. 1800. *Fig. 44.*

Cardamine laciniata (Muhl.) Wood, Bot. & Fl. 38. 1870.

Cardamine laciniata (Muhl.) Wood var. *integra* Schulz, Bot. Jahrb. 32:349. 1903.

Dentaria laciniata Muhl. var. *integra* (Schulz) Fern. Rhodora 10:84. 1908.

Herb from yellow-brown, fusiform tubers; stem slender, to 35 cm tall, glabrous or rarely pubescent; cauline leaves mostly 3, whorled or nearly so, palmately 3-parted nearly to the base, rarely unlobed, the divisions linear to oblong, sharply toothed to entire, glabrous, on petioles up to half as long as the blades; basal leaves developing mostly after flowering, similar to the cauline leaves but with longer petioles; inflorescence a terminal raceme, the rachis hirsute; pedicels glabrous or pubescent, up to 2 cm long in fruit; sepals green, up to 1 cm long; petals white or purple or pink, to 2 cm long; siliques ascending, linear, long subulate-beaked, to 5 cm long; seeds several, oval, flat, without wings.

COMMON NAMES: Toothwort; Pepper-root.

HABITAT: Woods.

RANGE: Quebec to Minnesota, south to Nebraska, Louisiana, and Florida.

ILLINOIS DISTRIBUTION: Common throughout the state; in every county.

The toothwort is one of the commonest early spring wildflowers in Illinois, coming into bloom in late February in the extreme southern tip of the state.

The plants show a great range of variability in overall stature, degree of leaf-cutting, and color of flowers. Many of these variations have received names, but none is worthy of recognition.

The tubers have a peppery taste and were commonly used by pioneers in salads.

Toothwort blooms from early March through May.

44. Dentaria laciniata (Toothwort). *a.* Habit, X¾. *b.* Flower, X2½. *c.* Sepal,
X5. *d.* Petal, X3½. *e.* Fruit, X1½. *f.* Seed, X10.

2. Raphanus L. – Radish

Annual or biennial herbs from slender or thick roots; leaves lyrate, mostly basal, the cauline ones progressively reduced in size toward the tip of the stem; inflorescence racemose, the flowers showy, actinomorphic; sepals 4, free; petals 4, free; stamens 6; pistil one, the ovary superior, the style slender; fruit a silique, cylindrical, sometimes torulose, indehiscent, beaked, the seeds several.

Raphanus is a genus of six species, all native to Europe and Asia.

KEY TO THE SPECIES OF Raphanus IN ILLINOIS

1. Petals yellowish or white; fruit up to 6 mm thick _____
   _____1. R. raphinistrum
1. Petals pale purple; fruit 6–10 mm thick _____ 2. R. sativus

1. Raphanus raphinistrum L. Sp. Pl. 669. 1753. Fig. 45.

Annual or biennial herbs from slender roots; stems erect or ascending, to 75 cm tall, much branched, sparsely hispid; leaves lyrate, to 20 cm long, the terminal lobe much larger than the other lobes, the lobes crenate or dentate, rough-hispid, the uppermost leaves much smaller, less divided to unlobed; racemes terminal, loosely several-flowered; flowers to 2 cm broad, on pedicels up to 1.2 cm long; sepals green, linear to narrowly lanceolate, hispidulous, to 1 cm long; petals yellow, fading to white, with purple veins, to 2 cm long, broadly rounded and usually erose at the apex, clawed at the base; silique cylindric, to 3 cm long, to 6 mm thick, becoming constricted between the seeds, longitudinally grooved, glabrous, 2- to 10-seeded, the beak subulate-conic, the pedicels up to 1.6 cm long.

COMMON NAMES: Wild Radish; Jointed Charlock.

HABITAT: Waste places.

RANGE: Native of Eruope and Asia; adventive throughout much of North America, including the West Indies.

ILLINOIS DISTRIBUTION: Known only from Boone, Champaign, DeKalb, Du Page, Mason, McHenry, Peoria, and St. Clair counties.

The wild radish is an infrequently found weed of waste places. Most of the Illinois specimens were collected along roadsides.

The yellow flowers, which fade badly to white at maturity, are borne from June to August.

45. *Raphanus raphinistrum* (Wild Radish). *a.* Upper part of plant, X⅓. *b.* Habit (silhouette), X1/14. *c.* Flower, X1½. *d.* Sepal, X5. *e.* Petal, X3½.

Unlike the radish (*R. sativus*), the roots of this species do not become fleshy.

2. Raphanus sativus L. Sp. Pl. 669. 1753. *Fig. 46.*

Annual or biennial herbs from thickened roots; stems erect, to 65 cm tall, hispid; leaves lyrate, to 20 cm long, the terminal lobe much longer than the other ones, the lobes crenate or dentate, rough-hispid, the uppermost leaves much smaller, less divided to unlobed; racemes terminal, loosely several-flowered; flowers to 2 cm broad, on pedicels up to 1.2 cm long; sepals green, narrowly lanceolate, hispidulous, to 1 cm long; petals pink or pale purple, to 2 cm long, broadly rounded and usually erose at the apex, clawed at the base; silique thickly cylindric, to 1.5 cm long, 6–10 mm thick, not strongly constricted between the seeds, longitudinally grooved, glabrous, 2- to 3-seeded, the beak subulate-conic, the pedicels up to 1.5 cm long.

COMMON NAME: Radish.

HABITAT: Waste places.

RANGE: Native of Europe and Asia; rarely spontaneous in North America.

ILLINOIS DISTRIBUTION: Known from several scattered localities in the state.

The radish is commonly cultivated in gardens. Stray specimens may occasionally persist for a year or two.

The thickened roots and pale purple flowers readily distinguish it from *R. raphinistrum*.

The flowers appear from May to August.

3. *Hesperis* L. – Rocket

Biennial or perennial herbs from thickened roots; leaves alternate, serrate; inflorescence racemose, the flowers showy, actinomorphic; sepals 4, free; petals 4, free; stamens 6; pistil one, the ovary superior, the style slender; fruit a silique, cylindrical, keeled, dehiscent, not prominently beaked, the seeds unwinged, in one row in each locule.

Hesperis is a genus of about 25 species, all native to Europe and Asia.

Only the following species occurs in Illinois.

1. Hesperis matronalis L. Sp. Pl. 663. 1753. *Fig. 47.*

Biennial or perennial herbs from slightly thickened roots; stems

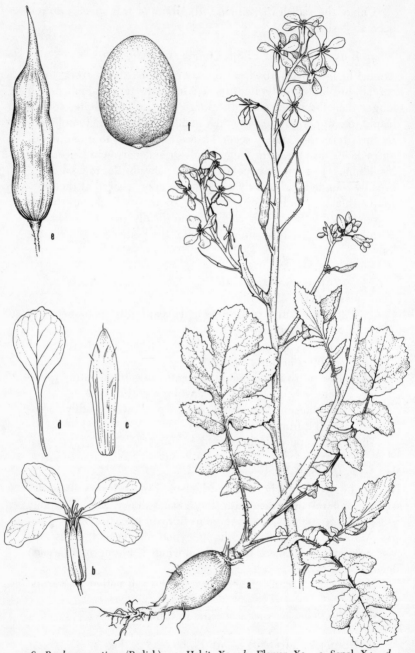

46. *Raphanus sativus* (Radish). *a.* Habit, X1. *b.* Flower, X2. *c.* Sepal, X4. *d.* Petal, X2. *e.* Fruit, X4. *f.* Seed, X10.

47. *Hesperis matronalis* (Dame's Rocket).   *a.* Habit, X¾.   *b.* Habit (silhouette), X1/36.   *c.* Flower, X1½.   *d.* Sepal, X4.   *e.* Petal, X3.   *f.* Fruit, X¾.   *g.* Seed, X10.

erect, unbranched or sparsely branched, to 1 m tall, hispidulous; leaves lanceolate to ovate-lanceolate, acute to acuminate at the apex, rounded at the base, or the lowermost leaves cuneate, to 20 cm long, the lower leaves nearly entire or denticulate, the upper leaves serrate, pubescent above and below, the petioles very short or absent; racemes terminal, loosely flowered; flowers fragrant, to 2.5 cm broad, on pedicels up to 3 cm long; sepals green, lanceolate, to 1 cm long, pubescent; petals usually purple, to 2.2 cm long, broadly rounded at the apex, tapering to a clawed base; silique narrowly cylindric, to 15 cm long, spreading to ascending, shallowly constricted between the seeds, glabrous, several-seeded, the seeds broadly elliptic, subulate-tipped, areolate.

COMMON NAMES: Dame's Rocket; Summer Lilac.

HABITAT: Waste ground, particularly in vacant lots.

RANGE: Native of Europe and Asia; sometimes persisting or escaping from cultivation in northeastern North America.

ILLINOIS DISTRIBUTION: Scattered throughout the state, but apparently more frequent in the northern half of Illinois.

Dame's rocket is a plant which in the past was often grown in flower gardens. It does not seem to be as popular today. Occasionally specimens which have escaped from cultivation may be found in waste ground.

The fragrant, purple flowers, which bloom from May to August, account for the common name of summer lilac.

4. *Iodanthus Torr. & Gray* – Purple Rocket

Perennial herbs from slender roots; leaves alternate, sometimes lyrate-pinnatifid, sometimes auriculate; inflorescence paniculate, the flowers actinomorphic, bractless; sepals 4, free; petals 4, free; stamens 6; pistil one, the ovary superior, the style stout; fruit a silique, cylindrical, dehiscent, the seeds unwinged, in one row in each locule.

Only the following species comprises the genus.

1. **Iodanthus pinnatifidus** (Michx.) Steud. Nom. Bot. ed. 2, 1:812. 1840. *Fig. 48.*

Hesperis pinnatifida Michx. Fl. Bor. Am. 2:31. 1803.

Cheiranthes hesperidoides Torr. & Gray, Fl. N. Am. 1:72. 1838.

Iodanthus hesperidoides (Torr. & Gray) Torr. & Gray ex Gray, Gen. Illus. 1:134. 1848.

48. Iodanthus pinnatifidus (Purple Rocket).   *a.* Habit, X¼.   *b.* Flower, X4.   *c.* Sepal, X7½.   *d.* Petal, X6.   *e.* Fruit, X4.

Arabis hesperidoides (Torr. & Gray) Gray, Man. Bot. ed. 5, 68. 1867.

Thelypodium pinnatifidum (Michx.) S. Wats. Bot. King's Expl. 5:25. 1871.

Stems slender, to 80 cm tall, branched in the upper portion, glabrous; basal leaves ovate to oblong, dentate, subacute to acute at the apex, usually cordate at the base, glabrous, on slender petioles; lower and middle cauline leaves ovate-oblong, subacute to acute at the apex, tapering to a winged petiole at the base, serrate, occasionally pinnatifid near base of blade, to 18 cm long, glabrous, the petiole usually auriculate at the base; upper cauline leaves similar but smaller and nearly sessile; panicles terminal or from the upper axils, the numerous flowers 5–9 mm broad, on glabrous, spreading pedicels up to 5 mm long; sepals 4 in 2 pairs, the inner pair slightly gibbous at the base, green, lanceolate, glabrous, about 3 mm long; petals 4, 6–12 mm long, including the 3–6 mm long claw, purplish; silique linear-cylindric, slightly flattened, spreading or ascending, to 3 cm long, glabrous, tipped by the persistent short style, several-seeded, the seeds oblong, wingless.

COMMON NAMES: Purple Rocket; Violet Cress.

HABITAT: Moist woods.

RANGE: Pennsylvania to Minnesota, south to Texas and Alabama.

ILLINOIS DISTRIBUTION: Scattered throughout the state, but not common.

Purple rocket is an infrequently occurring species of floodplain woods and moist wooded ravines.

It is similar to *Hesperis*, but differs in being completely glabrous and having smaller flowers.

The flowers are borne from May to July.

5. *Cakile* Mill. – Sea Rocket

Fleshy annual herbs from deep-seated roots; leaves alternate, toothed; inflorescence racemose, terminal, the flowers actinomorphic, bractless; sepals 4, free; petals 4, free; stamens 6; pistil one, the ovary superior; silicle 2-jointed, fleshy, each joint 1-seeded, indehiscent.

Cakile is a genus of about six species native to seashores and lakeshores in Europe and North America.

Only the following taxon occurs in Illinois.

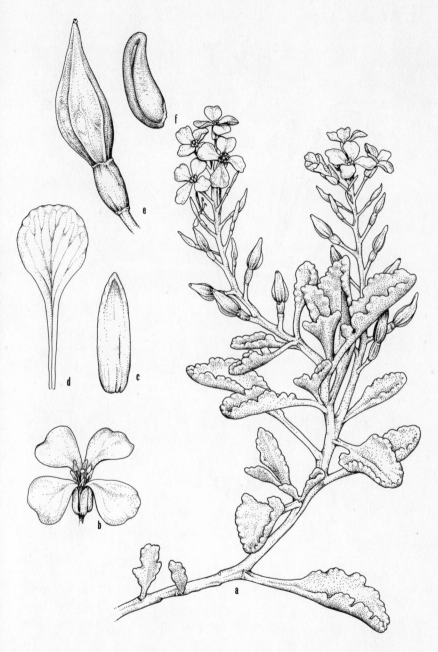

49. Cakile edentula (Sea Rocket). *a.* Upper part of plant, X1½. *b.* Flower, X2½.
c. Sepal, X11. *d.* Petal, X10. *e.* Fruit, X10. *f.* Seed, X10.

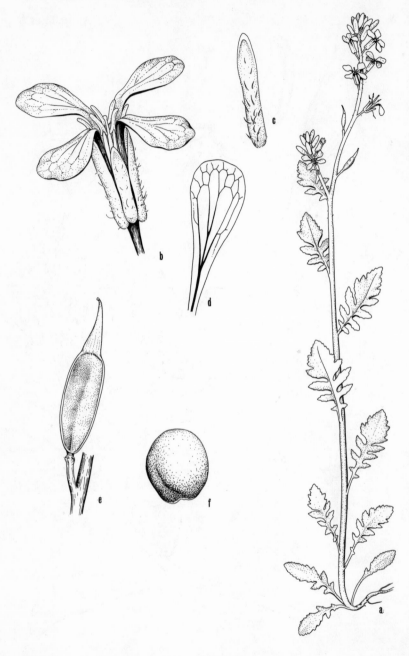

50. *Eruca sativa* (Garden Rocket).   *a.* Habit, X¼.   *b.* Flower, X4.   *c.* Sepal, X6.
d. Petal, X4.   *e.* Fruit, X5.   *f.* Seed, X10.

1. **Cakile edentula** (Bigel.) Hook. var. **lacustris** Fern. Rhodora 24:23. 1922. *Fig. 49.*

Fleshy annual; stems much branched, fleshy, glabrous, to 75 cm long; leaves obovate to oblanceolate, obtuse at the apex, cuneate at the base, sinuate-dentate, to 8 cm long, glabrous; racemes terminal, the flowers 4–6 mm broad, on very short, stout, glabrous pedicels; sepals 4, elliptic, green, glabrous, to 3 mm long; petals 4, pale purple, 4–6 mm long, long-clawed; silicle 2-jointed, the upper ovoid-lanceolate, flattened, 4–6 mm long, glabrous, with one erect seed, the lower obovoid, terete, 2–4 mm long, glabrous, with one suspended seed or seedless.

COMMON NAME: Sea Rocket.

HABITAT: Beaches of Lake Michigan.

RANGE: States bordering the Great Lakes.

ILLINOIS DISTRIBUTION: Cook and Lake counties.

I am following Fernald (1922) in calling our plants var. *lacustris*, which differs from var. *edentula* primarily by its longer beaked silicle.

The fleshy nature of the sea rocket makes it a striking species in the sandy beaches along Lake Michigan.

The first Illinois collection of this plant was made in 1860 by Franklin Scammon, an early Chicago physician. Until Fernald segregated var. *lacustris* in 1922, our plants were known mostly as *Cakile americana* Nutt.

The flowers open from July to September.

6. *Eruca Mill.* – Garden Rocket

Annual or biennial, sometimes succulent, herbs; leaves alternate, pinnatifid; inflorescence racemose; flowers actinomorphic, bractless; sepals 4, free; petals 4, free; stamens 6; pistil one, the ovary superior, the style elongated; siliques linear-oblong, beaked, dehiscent, the valves 3-nerved, the seeds borne in two rows in each locule.

About ten European and Asian species comprise the genus.

Only the following species occurs in Illinois.

1. **Eruca sativa** Mill. Gard. Dict. ed. 8, no. 1. 1768. *Fig. 50.*

Brassica eruca L. Sp. Pl. 667. 1753.

Eruca eruca (L.) Britt. in Britt. & Brown, Ill. Fl. 2:192. 1913.

Rather succulent annual; stems much branched, glabrous, to 1.3 m tall; leaves pinnatifid, obtuse to subacute at the apex, narrowed to a winged petiole, to 15 cm long, glabrous, the uppermost leaves smaller and less deeply lobed or sometimes merely dentate; racemes terminal or from the upper axils, the flowers 15–20 mm broad, on stout, glabrous pedicels up to 5 mm long; sepals 4, green; petals 4, obovate, yellow or purplish with violet veins; silique erect, fusiform, more or less 4-angled, 10–15 mm long, glabrous, with a stout, flat beak, with several ellipsoid seeds.

COMMON NAME: Garden Rocket.
HABITAT: Cultivated field.
RANGE: Native of Europe; occasionally adventive in much of the United States.
ILLINOIS DISTRIBUTION: Known only from Peoria County. The garden rocket at one time was cultivated as a food plant for salads because of the peppery taste of the leaves. The only Illinois collection known from plants not in cultivation was made by F. E. McDonald in Peoria County on July 13, 1907.

The flowers appear from June to September.

7. *Capsella* Medic. – Shepherd's-purse

Annual herbs from an elongated root and with branched hairs; leaves basal and cauline, alternate, sometimes pinnatifid; inflorescence racemose; flowers actinomorphic, bractless; sepals 4, free; petals 4, free; stamens 6; pistil one, the ovary superior, the style short; silicles obcordate-triangular, compressed at right angles to the septum, the valves keeled, the seeds numerous.

Capsella is a genus of about four species native to Europe.

Only the following species occurs in Illinois.

1. **Capsella bursa-pastoris** (L.) Medic. Pfl. Gatt. 1:85. 1792.
Fig. 51.

Thlaspi bursa-pastoris L. Sp. Pl. 647. 1753.

Bursa bursa-pastoris (L.) Britt. Mem. Torrey Club 5:172. 1894.

Winter annual with an elongated taproot; stems erect, to 60 cm tall, often branched, hirsute to nearly glabrous; rosette leaves often runcinate-pinnatifid but variously lobed or toothed, acute at the apex, tapering to a winged petiole, more or less pubescent, to 12 cm long; cauline leaves lanceolate, acute at the apex, auriculate at the sessile base, dentate to entire, more or less pubescent; racemes mostly terminal, the flowers to 3 mm broad, on spreading to ascending

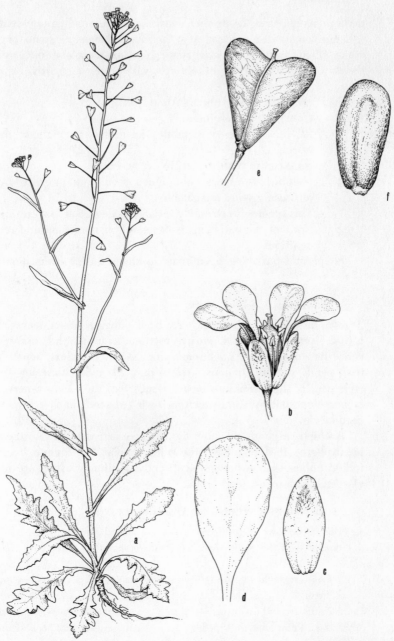

51. *Capsella bursa-pastoris* (Shepherd's-purse). *a*. Habit, X¼. *b*. Flower, X10.
c. Sepal, X25. *d*. Petal, X15. *e*. Fruit, X4. *f*. Seed, X10.

pedicels longer than the flowers; sepals 4, ovate, usually pubescent, 1–2 mm long; petals 4, broadest at the rounded apex, 2–4 mm long, white; silicles triangular, emarginate, 5–10 mm long, glabrous, with the short persistent style in the notch, with numerous seeds.

COMMON NAME: Shepherd's-purse.

HABITAT: Waste ground.

RANGE: Native of Europe; adventive throughout the world.

ILLINOIS DISTRIBUTION: In every county.

Shepherd's-purse is a ubiquitous weed found in every kind of waste ground imaginable.

This species is extremely variable in degree of leaf cutting. None of the variations is deserving of nomenclatural recognition.

Shepherd's-purse has been found in bloom during every month of the year.

8. *Arabis* L. – Rock Cress

Annual, biennial, or perennial herbs from elongated roots; leaves alternate, sometimes basal, entire, toothed, or pinnatifid; inflorescence racemose; flowers actinomorphic, mostly bractless; sepals 4, free; petals 4, free; stamens 6; pistil one, the ovary superior, the style usually short; siliques linear, usually flat, the valves 1-nerved or nerveless, with several sometimes winged seeds in 1–2 rows in each locule.

Arabis is a genus of about 125 species native to the northern hemisphere. It differs from the often similar *Cardamine* by its forked pubescence (when present) and its siliques which are not elastically dehiscent.

KEY TO THE SPECIES OF Arabis IN ILLINOIS

1. None of the leaves sagittate nor auriculate at the base _____ 2
1. Some or all of the cauline leaves sagittate or auriculate at the base ____
   --3
   2. Fruits strongly ascending at maturity; leaves on stem up to 5 mm broad _____1. A. *lyrata*
   2. Fruits arching and usually pointing downward at maturity; leaves on stem 5 mm broad or broader _____ 2. A. *canadensis*
3. Stems glabrous (except sometimes at the very base in A. *glabra* and A. *drummondii*) _____ 4
3. Stems hairy to summit _____ 6

4. At least some of the fruits over 1 mm wide; petals white _____ 5
4. Fruit never exceeding 1 mm in width; petals creamy-yellow _____
_____3. *A. glabra*
5. Petals about as long as the sepals; siliques ascending to spreading; seeds arranged in one row _____ 5. *A. laevigata*
5. Petals about twice as long as the sepals; siliques erect, appressed; seeds arranged in two rows _____ 4. *A. drummondii*
6. Leaves and stems with simple and stellate hairs; petals 2–3 mm long; seeds wingless _____ 6. *A. shortii*
6. Leaves and stems with simple hairs; petals 4–6 mm long; seeds narrowly winged _____ 7. *A. hirsuta*

1. **Arabis lyrata** L. Sp. Pl. 665. 1753. *Fig. 52.*
Cardamine spathulata Michx. Fl. Bor. Am. 2:29. 1803.

Tufted perennial from a taproot; stems slender, erect, to 35 cm tall, usually branched from the base, glabrous or usually hirsute, at least in the lower half; basal leaves lyrate-pinnatifid, rarely merely dentate, spatulate to oblanceolate, to 4 cm long, more or less hirsute on both surfaces; cauline leaves spatulate to linear, obtuse to subacute at the apex, cuneate at the base, entire or dentate, rarely pinnatifid, usually hirsute, at least below, rarely completely glabrous, to 2.5 cm long, to 5 mm broad; racemes terminal, the flowers 6–9 mm broad, on slender, ascending pedicels to nearly 1 cm long; sepals 4, broadly elliptic, glabrous, 1.5–3.0 mm long, green; petals 4, spatulate to oblanceolate, 5–8 mm long, white; siliques linear, ascending, more or less flattened, glabrous, to 4.5 cm long, up to 2 mm broad, the valves 1-nerved to above the middle, the persistent style about 1 mm long, the seeds oblong, wingless, about 1 mm long, arranged in a single row.

COMMON NAMES: Sand Cress; Lyre-leaved Rock Cress.
HABITAT: Sand dunes; sandy woods; dry, gravelly prairies.
RANGE: Vermont to Minnesota, south to Missouri and Georgia; Ontario and Alberta.
ILLINOIS DISTRIBUTION: Restricted to the northern one-fourth of Illinois.
Although this species is generally not common, it is abundant on the sand dunes near Lake Michigan.
It differs from all other *Arabis* species in Illinois except *A. canadensis* by the absence of auricles at the base of the leaves. From *A. canadensis* it differs by its ascending siliques and its narrow cauline leaves.

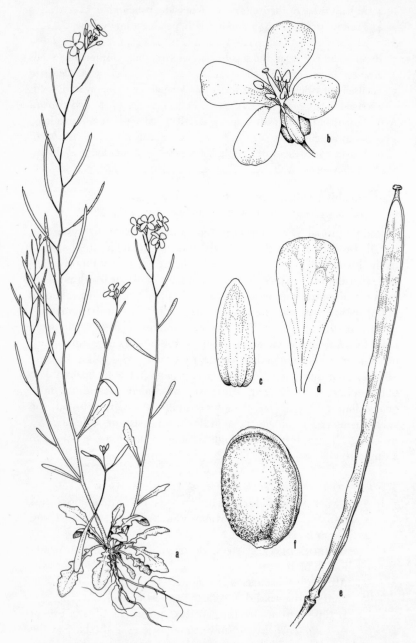

52. *Arabis lyrata* (Sand Cress). *a.* Habit, X½. *b.* Flower, X5. *c.* Sepal, X12½.
d. Petal, X6. *e.* Fruit, X4. *f.* Seed, X20.

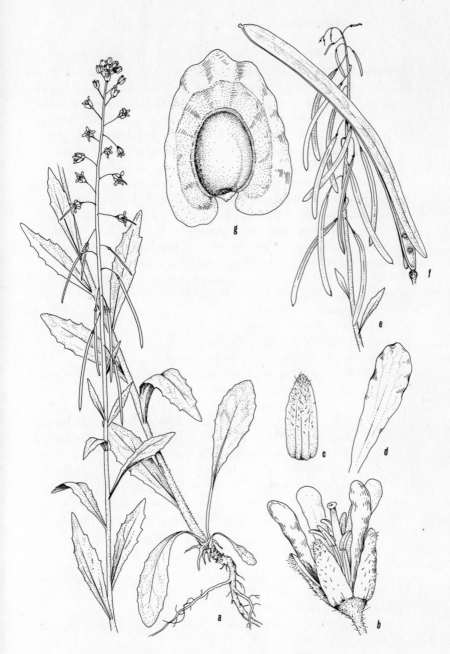

53. *Arabis canadensis* (Sicklepod). *a.* Habit, X½. *b.* Flower, X7½. *c.* Sepal, 8½. *d.* Petal, X8½. *e.* Fruiting raceme, X⅓. *f.* Fruit, X1. g. Seed, X37½.

The sand cress flowers from late May through July.

2. Arabis canadensis L. Sp. Pl. 665. 1753. *Fig. 53.*

Biennial from an elongated taproot; stems erect, to 75 cm tall, usually unbranched, hirsute near the base, glabrous above; basal leaves obovate to lanceolate, obtuse to subacute at the apex, cuneate at the base, lyrate or merely dentate, to 15 cm long, to 4 cm broad, hirsute at least along the midvein on both surfaces, withering early in the summer; cauline leaves oblong to elliptic, acute at the apex, cuneate at the sessile base, denticulate, to 12 cm long, to 2.5 cm broad, the lowest villous-hirsute, the upper less pubescent to nearly glabrous; racemes terminal, the flowers loosely arranged, 4–7 mm broad, on pendulous pubescent pedicels to nearly 1 cm long; sepals 4, lanceolate, glabrate to pubescent, green, 2.0–3.5 mm long; petals 4, narrowly oblong to spatulate, white, 3–5 mm long; siliques strongly falcate, recurved or at length pendulous, to 10 cm long, to 5 mm broad, glabrous, reticulate, the valves 1-nerved nearly to the tip, the persistent style minute, the seeds oblong, winged, 1.0–1.3 mm long, arranged in a single row.

COMMON NAME: Sicklepod.

HABITAT: Dry, often rocky, woods.

RANGE: Maine to Minnesota, south to Texas and Georgia; also southern Ontario.

ILLINOIS DISTRIBUTION: Occasional throughout the state. The sicklepod is distinctive because of its strongly falcate siliques and the lack of any auriculate leaves. The valves of the fruit are long persistent on the plant, often remaining until after all the leaves have withered.

In the southern half of the state, sicklepod occurs mostly in dry, rocky woods. In the area near Lake Michigan, however, Swink (1974) attributes it to the shade of dune woods, on sloping ground.

This species flowers from May to July.

3. Arabis glabra (L.) Bernh. Verz. Syst. Erf. 195. 1800. *Fig. 54.*
Turritis glabra L. Sp. Pl. 666. 1753.
Arabis perfoliata Lam. Encycl. 1:219. 1783.
Arabis glabra (L.) Bernh. var. *typica* Hopkins, Rhodora 39:106. 1937.

Biennial from an elongated taproot; stems erect, rather stout, to 1.2 m tall, usually unbranched, hirsute at the base, glabrous and glau-

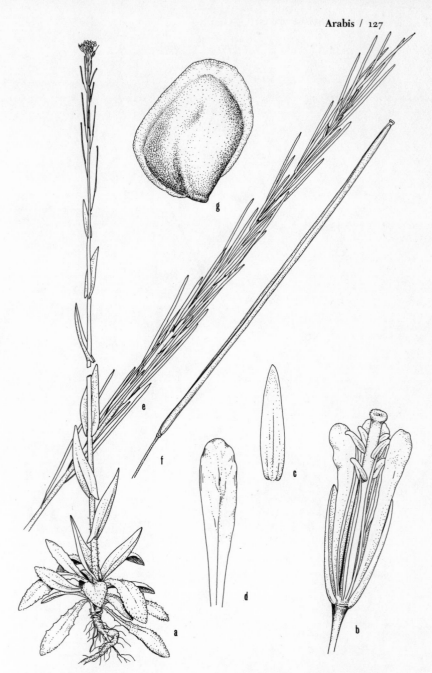

54. *Arabis glabra* (Tower Mustard). *a.* Habit, X1/6. *b.* Flower (with one sepal and one petal removed), X10. *c.* Sepal, X10. *d.* Petal, X10. *e.* Fruiting raceme, X⅓. *f.* Fruit, X2. *g.* Seed, X37½.

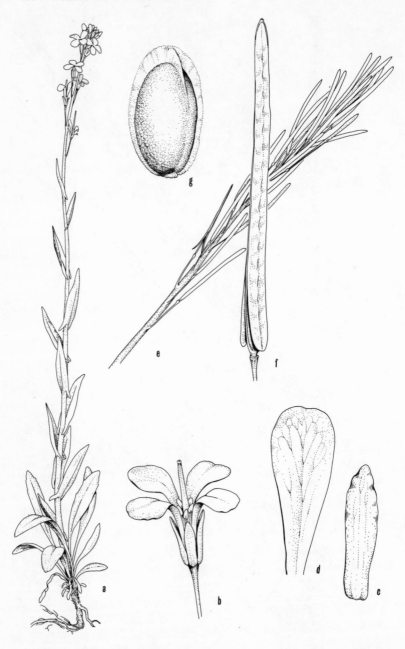

55. *Arabis drummondii* (Drummond's Rock Cress). *a.* Habit, X¼. *b.* Flower, X5. *c.* Sepal, X10. *d.* Petal, X6. *e.* Fruiting raceme, X¼. *f.* Fruit, X1½. *g.* Seed, X35.

cous above; basal leaves oblong to oblanceolate, subacute at the apex, cuneate at the base, dentate, less commonly lyrate or entire, to 12 cm long, stellate-pubescent when young; cauline leaves oblong to lanceolate, subacute to acute at the apex, sagittate at the sessile base, entire, to 12 cm long, glabrous or nearly so; racemes terminal, virgate, the flowers 4–5 mm broad, on slender, erect, glabrous pedicels to 1.5 cm long; sepals 4, lanceolate, glabrous, green, 2–4 mm long; petals 4, narrowly oblanceolate to linear, yellowish-white, 3–6 mm long; siliques straight, erect and appressed, linear, terete, to 8 cm long, about 1 mm broad, glabrous, the valves 1-nerved at least to the middle, the persistent style minute, the seeds narrowly winged, about 1 mm long, arranged in 1–2 rows.

COMMON NAME: Tower Mustard.

HABITAT: Moist prairies; limestone woods.

RANGE: Quebec to Alaska, south to California, Arkansas, and North Carolina; Europe; Asia.

ILLINOIS DISTRIBUTION: Occasional in the northern half of the state, rare in the southern half.

This species is similar to *A. hirsuta* and *A. drummondii* because of the erect, appressed siliques. *Arabis glabra* differs by its longer, somewhat terete siliques and its very short, thick style.

Early Illinois botanists such as Patterson (1876) and Brendel (1887) called this species *A. perfoliata* Lam.

The tower mustard blooms from May to July.

4. Arabis drummondii Gray, Proc. Am. Acad. 6:187. 1866. *Fig. 55.*

Biennial from a slender taproot; stems erect, rather stout, to 1 m tall, branched or unbranched, glabrous or hirsutulous at the base, usually more or less glaucous; basal leaves oblanceolate to obovate, obtuse to subacute at the apex, cuneate at the base, dentate to entire, to 9 cm long, pilose when young, becoming glabrous, the petioles to 4 cm long, glabrous or ciliate; cauline leaves lanceolate to oblong, subacute at the apex, sagittate at the base, serrate to entire, glabrous at maturity, to 7 cm long; racemes terminal, the flowers 6–8 mm broad, on rather stout, erect, glabrous pedicels; sepals 4, lance-elliptic, glabrous, green, 2.5–4.5 mm long; petals 4, oblanceolate to spatulate, pink or whitish, 5–10 mm long; siliques straight, erect, appressed, flat, to 10 cm long, to 3.5 mm broad, glabrous, the persistent style short, the seeds spherical, narrowly winged, about 1 mm in diameter, arranged in 2 rows.

COMMON NAME: Drummond's Rock Cress.

HABITAT: Gravelly soil.

RANGE: Labrador to British Columbia, south to California, Arizona, New Mexico, eastern Iowa, northeastern Illinois, northern Ohio, and Delaware.

ILLINOIS DISTRIBUTION: Known only from Cook and Kane counties.

This rare rock cress has not been found in Illinois since late in the nineteenth century. It is very similar to *A. divaricarpa* Nels. (= *A. brachycarpa* [Torr. & Gray] Britt.), with which it is often confused, by its erect, appressed siliques and its sometimes hirsutulous stems. It differs from *A. hirsuta* by having its seeds in two rows.

Hopkins's (1937) report of a large colony of *A. drummondii* in Champaign County apparently is erroneous.

Although our plant has been referred to by Jones et al. (1955) as *A. confinis* S. Wats., this binomial properly belongs as a synonym under *A. brachycarpa* (Torr. & Gray) Britt.

The Drummond's rock cress flowers from late May to July.

5. **Arabis laevigata** (Muhl.) Poir, in Lam. Encycl. Suppl. 1:411. 1810. *Fig. 56.*

Turritis laevigata Muhl. ex Willd. Sp. Pl. 3:543. 1801.

Biennial from a slender taproot; stems erect, to 1 m tall, branched or unbranched, glabrous, glaucous; basal leaves spatulate to obovate, subacute to acute at the apex, cuneate to the petiolate base, serrate, to 10 cm long, to 3 cm broad, pilose when young, becoming glabrous, usually absent at flowering times; cauline leaves linear to lanceolate, acute at the apex, sagittate at the base, entire or serrate, to 20 cm long, pilose when young, becoming glabrous; racemes terminal, loosely flowered, the flowers 4–7 mm broad, on slender, ascending, glabrous pedicels to 1 cm long; sepals 4, lance-elliptic, glabrous, green, 2.0–3.5 mm long; petals 4, white, 3–5 mm long; siliques straight or arcuate, more or less terete, to 10 cm long, to 2.5 mm broad, glabrous, the style minute, the seeds oblong, broadly winged, about 1 mm long, arranged in one row.

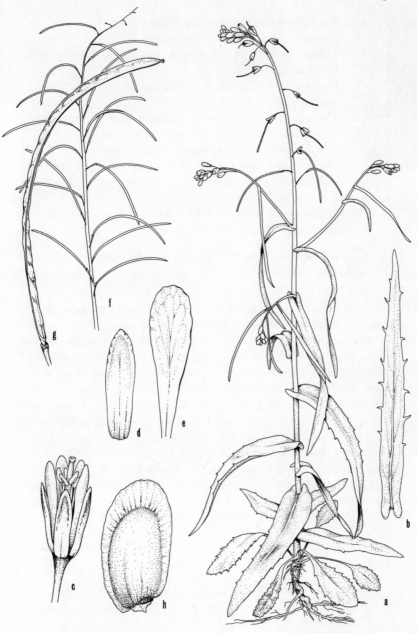

56. Arabis laevigata (Smooth Rock Cress). *a*. Habit, X½. *b*. Leaf, X¾. *c*. Flower, X5. *d*. Sepal, X10. *e*. Petal, X10. *f*. Fruiting raceme, X¼. *g*. Fruit, X1. *h*. Seed, X37½.

COMMON NAME: Smooth Rock Cress.

HABITAT: Moist, often rocky, woods.

RANGE: Quebec to Minnesota, south to Colorado, Oklahoma, and Georgia.

ILLINOIS DISTRIBUTION: Occasional to common throughout the state.

The smooth rock cress is distinguished by its sagittate leaves, glabrous stems, and arcuate siliques.

Considerable variation exists in the shape of the leaves and the amount of serrations along the margins. The basal leaves, which usually overwinter, are purple in color on the lower surface.

The most frequent habitat for this species is in rocky woods.

The flowers appear from March to June.

6. Arabis shortii (Fern.) Gl. Phytologia 4:23. 1952.

Arabis perstellata E. L. Braun var. *shortii* Fern. Rhodora 48: 208. 1946.

Biennial from a slender taproot; stems erect, to 75 cm tall, usually sparsely branched, stellate-pubescent; basal leaves obovate, subacute to acute at the apex, cuneate to the petiolate base, dentate, to 15 cm long, to 6 cm broad, usually strigose above, stellate-pubescent beneath; cauline leaves lanceolate to oblong, obtuse to acute at the apex, auriculate at the base, dentate, to 6 cm long, glabrous or strigose on the upper surface, stellate-pubescent below; raceme terminal, the flowers 2.5–4.5 mm broad, on ascending, hirsute pedicels about as long as the flower; sepals 4, elliptic, more or less pubescent, green, 1.5–2.5 mm long; petals 4, white, 2–3 mm long; siliques nearly straight, spreading to ascending, linear, to 4 cm long, to 1.5 mm broad, stellate-pubescent or glabrous, the style minute, the seeds oblong, wingless, about 1 mm long, arranged in one row.

Two varieties may be distinguished in Illinois.

1. Siliques stellate-pubescent _____ 6a. *A. shortii* var. *shortii*
1. Siliques glabrous _____ 6b. *A. shortii* var. *phalacrocarpa*

6a. Arabis shortii (Fern.) Gl. var. **shortii** *Fig. 57a–e.*

Arabis dentata Torr. & Gray, Fl. N. Am. 1:80. 1838, non de Clairville (1811).

Arabis perstellata E. L. Braun var. *shortii* Fern. Rhodora 48: 208. 1946.

Siliques stellate-pubescent.

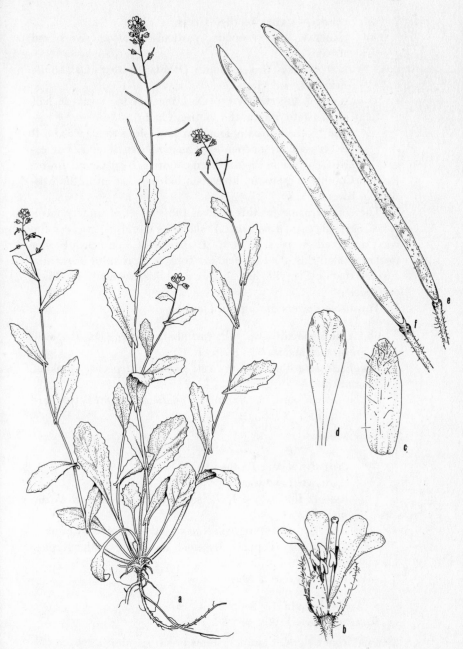

57. *Arabis shortii* (Toothed Cress). *a.* Habit, X¼. *b.* Flower, X7½. *c.* Sepal,
X15. *d.* Petal, X15. *e.* Fruit, X2½. var. *phalacrocarpa.* *f.* Fruit, X2½.

COMMON NAME: Toothed Cress.

HABITAT: Rocky woods, particularly along rivers and streams.

RANGE: New York to South Dakota, south to Oklahoma, Arkansas, and Virginia.

ILLINOIS DISTRIBUTION: Occasional in the northern half of the state, rare in the southern half.

The toothed cress is found in low, often rocky woods. It is occasionally encountered in northern Illinois, but extremely rare in the southern counties. Near Grand Tower, Jackson County, it grows on limestone ledges adjacent to the Mississippi River.

The correct name for this plant is the subject of much controversy. Fernald (1950) insists that *A. shortii* is merely a variety of the very restricted *A. perstellata* E. L. Braun of Kentucky. Earlier workers called this plant *A. dentata* Torr. & Gray until it was discovered that de Clairville had already used this binomial for a different species.

This taxon flowers during April and May.

6b. Arabis shortii (Fern.) Gl. var. **phalacrocarpa** (M. Hopkins) Steyerm. Rhodora 62:130. 1960. *Fig. 57f.*

Arabis dentata Torr. & Gray var. *phalacrocarpa* M. Hopkins, Rhodora 39:169. 1937.

Arabis perstellata E. L. Braun var. *phalacrocarpa* (M. Hopkins) Fern. Rhodora 48:208. 1946.

Siliques glabrous.

COMMON NAME: Toothed Cress.

HABITAT: Low woods.

RANGE: Illinois and Iowa, south to Oklahoma and Arkansas.

ILLINOIS DISTRIBUTION: Known only from Lake County.

This variety is apparently much less common than typical var. *shortii* in Illinois.

It flowers during May.

7. Arabis hirsuta (L.) Scop. Fl. Carn. ed. 2, 2:30. 1772.

Turritis hirsuta L. Sp. Pl. 666. 1753.

Biennial from a slender taproot; stems erect, slender, to 75 cm tall,

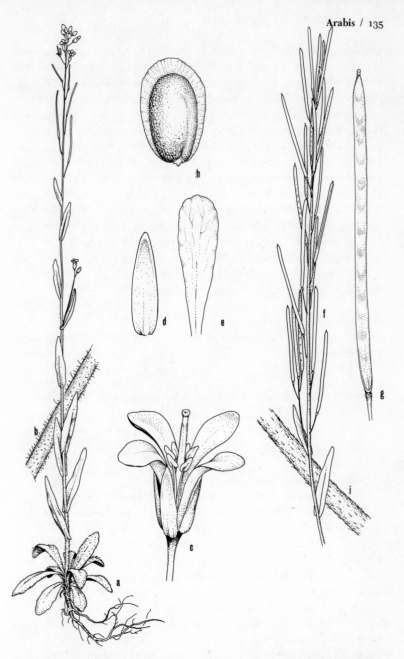

58. *Arabis hirsuta* (Hairy Rock Cress). var. *pycnocarpa*. *a*. Habit, X¼. *b*. Stem section, X2½. *c*. Flower, X5. *d*. Sepal, X7½. *e*. Petal, X7½. var. *adpressipilis*. *f*. Fruiting raceme, X⅓. *g*. Fruit, X2½. *h*. Seed, X37½. *i*. Stem section, X2½.

usually unbranched, strigose to hirsute or sometimes glabrous above, the hairs simple or forked; basal leaves oblong to oblanceolate, obtuse to subacute at the apex, cuneate to the petiolate base, dentate, to 8 cm long, villous-hirsute to strigose on both surfaces; cauline leaves lanceolate to oblong, obtuse to subacute at the apex, the lowest auriculate at the base, the uppermost merely sessile, dentate, to 4 cm long, villous-hirsute to strigose to nearly glabrous; racemes terminal, the flowers 4–6 mm broad, on erect, pubescent or glabrous pedicels; sepals 4, elliptic to lanceolate, pubescent or glabrous, green, 3–5 mm long; petals 4, white, 4–6 mm long; siliques nearly straight, erect and often appressed, linear, flat, to 5 cm long, up to 1 mm broad, glabrous, the style minute, the seeds oblong, winged, 1.0–1.3 mm long, arranged in one row.

Typical var. *hirsuta* is a European and Asian plant with plump siliques and more broadly winged seeds.

Three varieties may be found in Illinois.

1. Stems pubescent to summit _____ 2
1. Stems glabrous in the upper half _____ 7c. *A. hirsuta* var. *glabrata*
  2. Stems hirsute _____ 7a. *A. hirsuta* var. *pycnocarpa*
  2. Stems strigose _____ 7b. *A. hirsuta* var. *adpressipilis*

7a. Arabis hirsuta (L.) Scop. var. **pycnocarpa** (M. Hopkins) Rollins, Rhodora 43:318. 1941. *Fig. 58a–e.*
Arabis pycnocarpa M. Hopkins, Rhodora 39:112. 1937.

Stems hirsute to summit, the hairs mostly simple.

COMMON NAME: Hairy Rock Cress.
HABITAT: Limestone cliffs; woods; gravel prairies.
RANGE: Quebec to Alaska, south to California, New Mexico, Missouri, and Georgia.
ILLINOIS DISTRIBUTION: Occasional throughout the state. This is the more often found variety in Illinois, but it, too, is relatively rare. It is primarily a plant of limestone ravines, although Swink (1974) describes a station in Will County in a disturbed gravelly prairie.

Those persons considering this taxon specifically distinct from *A. hirsuta* call it *A. pycnocarpa* M. Hopkins.

Englemann wrote of the occurrence of this plant in Illinois as early as 1843.

7b. Arabis hirsuta (L.) Scop. var. **adpressipilis** (M. Hopkins) Rollins, Rhodora 43:319. 1941. *Fig. 58f–i.*
Arabis pycnocarpa M. Hopkins var. *adpressipilis* M. Hopkins, Rhodora 39:117. 1937.

Stems strigose to summit, the hairs mostly bifurcate.

HABITAT: Limestone woods.
RANGE: Ontario and Minnesota, south to Missouri and Virginia.
ILLINOIS DISTRIBUTION: Known only from Cook, Kane, St. Clair, and Will counties.

7c. Arabis hirsuta (L.) Scop. var. **glabrata** Torr. & Gray, Fl. N. Am. 1:80. 1838.

Stems glabrous above the middle.

HABITAT: Woods.
RANGE: Alberta to British Columbia, south to California and New Mexico; southwestern Wisconsin; northeastern Illinois.
ILLINOIS DISTRIBUTION: Known from Kankakee County.

9. *Descurainia Webb & Berthlelot* – Tansy Mustard

Annual or perennial herbs, mostly with bifurcate or stellate hairs; leaves bi- or tripinnate or -pinnatifid; inflorescence in terminal racemes; sepals 4, free, caducous; petals 4, free, yellow; stamens 6; pistil one, the ovary superior, the style minute; siliques linear-cylindric, the valves 1-nerved, the seeds very small, wingless, in one or two rows in each cell.

Descurainia is a genus of about two dozen species native to the northern hemisphere, the Andes Mountains of South America, and the Canary Islands. A number of species are native to the western United States. Early workers often used the name *Sophia* for this genus.

KEY TO THE TAXA OF Descurainia IN ILLINOIS

1. Siliques up to 1 mm thick; stems and leaves grayish, not glandular-pubescent _____1. D. sophia
1. Siliques 1–2 mm thick; stems and leaves green, somewhat glandular-pubescent _____2. D. pinnata var. brachycarpa

1. **Descurainia sophia** (L.) Webb in Engler & Prantl, Nat. Pflanzenf. 3(2):192. 1892. *Fig. 59.*
Sisymbrium sophia L. Sp. Pl. 659. 1753.
Sophia sophia Britt. in Britt. & Brown, Ill. Fl. 2:144. 1897.

Annual herb; stems bushy branched, to 75 cm tall, stellate-pubescent, gray; leaves bi- to tripinnate, the segments linear to linear-oblong, stellate-pubescent, grayish, the basal leaves withered by flowering time; racemes terminal, the numerous flowers 4–6 mm broad, on nearly glabrous, ascending pedicels to 1.5 cm long; sepals broadly elliptic, 2.0–2.5 mm long, about as long as the petals, green, glabrous or ciliate; petals greenish-yellow, obovate to oblanceolate, 2–3 mm long; siliques arcuate, ascending, linear-cylindric, to 3 cm long, to 1 mm thick, with 10–20 seeds per locule, the seeds oblong-ellipsoid, 0.7–1.0 mm long, arranged in one row.

COMMON NAMES: Flixweed; Tansy Mustard.
HABITAT: Waste areas.
RANGE: Native of Europe; occasionally adventive throughout the United States.
ILLINOIS DISTRIBUTION: Not common in the northern half of the state, rarer southward.
Descurainia sophia is distinguished from the other *Descurainia* in Illinois by its gray stems and leaves and its narrower siliques.
The flixweed is found principally along roads, railroads, and in barnyards. It flowers from June to August.

2. **Descurainia pinnata** (Walt.) Britt. var. **brachycarpa** (Richards.) Fern. Rhodora 42:266. 1940. *Fig. 60.*
Sisymbrium brachycarpum Richards. Frankl. Journ. App. 744. 1823.
Sophia intermedia Rydb. Mem. N. Y. Bot. Gard. 1:184. 1900.
Descurainia intermedia (Rydb.) Daniels, Univ. Mo. Stud., Sci. Ser. 1(2):147. 1907.
Descurainia brachycarpa (Richards.) O. E. Schulz, Pflanzenr. 105:325. 1924.

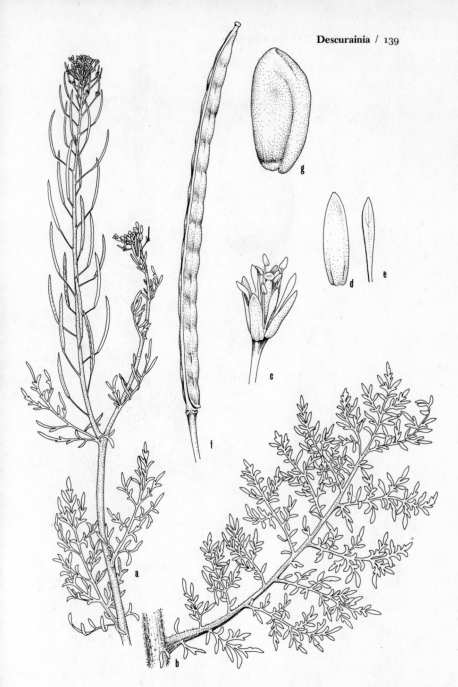

59. *Descurainia sophia* (Flixweed). *a.* Upper part of plant, X½. *b.* Stem and leaf from lower part of plant, X1¼. *c.* Flower, X6. *d.* Sepal, X10. *e.* Petal, X10. *f.* Fruit, X4. *g.* Seed, X37½.

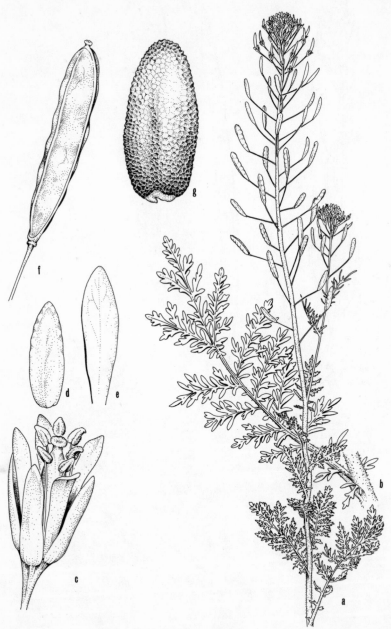

60. *Descurainia pinnata* (Tansy Mustard). *a.* Upper part of plant, X½. *b.* Stem and leaf from lower part of plant, X1¼. *c.* Flower, X17½. *d.* Sepal, X17½. *e.* Petal, X17½. *f.* Fruit, X6. *g.* Seed, X37½.

Annual herb; stems erect, branched, to 70 cm tall, glandular-pubescent, green; leaves bi- or tripinnate or bipinnatifid, the segments linear-oblong, more or less glandular-pubescent, green, the basal leaves withered by flowering time; racemes terminal, the numerous flowers 2–3 mm broad, on spreading-ascending pedicels to nearly 2 cm long; sepals 4, lanceolate, glabrous or nearly so, green, 1.0–1.5 mm long; petals 4, pale yellow, 1.5–3.5 mm long; siliques clavate, spreading-ascending, to 1 cm long, 1–2 mm thick, with 4–10 seeds per locule, the seeds oblong-ellipsoid, 0.7–1.0 mm long, arranged in two rows.

COMMON NAME: Tansy Mustard.

HABITAT: Along railroads; sandstone and limestone cliffs.

RANGE: Quebec to Mackenzie, south to Colorado, Texas, and Florida.

ILLINOIS DISTRIBUTION: Occasional throughout the state. Although this species frequently grows along railroads like an adventive, it also can be found on sandstone and limestone ledges.

Typical var. *pinnata*, which usually lacks glandular-pubescence and which has its siliques often horizontally spreading, occurs in southeastern Missouri but has not as yet been found in Illinois.

Variety *brachycarpa* flowers from April to June.

10. *Cardamine* L. – Bitter Cress

Perennial (less commonly annual) herbs from bulbous rootstocks or fibrous roots; leaves cauline and basal, simple or pinnately divided; inflorescence mostly racemose, with white or purple flowers; sepals 4, free; petals 4, free; stamens (4–) 6; pistil one, the ovary superior, the style small; siliques linear, flat, dehiscing elastically from the base, the valves essentially nerveless, the seeds compressed, wingless, arranged in one row in each cell.

Cardamine is a genus of about 125 species native in most temperate regions of the world. Although most of the species have small flowers, a few of them have rather showy flowers. One such pretty species, the cuckoo-flower (*C. pratensis* L.), is sometimes grown as a garden ornamental. The fresh stems and leaves of some species are eaten in salads.

KEY TO THE TAXA OF Cardamine IN ILLINOIS

1. None of the leaves pinnate or pinnatifid ----------------------- 2

1. Some or all the leaves pinnate or pinnatifid ---------------------- 3
   2. Stems glabrous; sepals green --------------------- 1. *C. bulbosa*
   2. Stems hirsute; sepals purplish ------------------ 2. *C. douglassii*
3. Flowers 10–15 mm broad; sepals 3–5 mm long; petals 8–15 mm long
   -----------------------------------3. *C. pratensis* var. *palustris*
3. Flowers 3–5 mm broad; sepals 0.5–2.0 mm long; petals 1.5–4.0 mm
   long --4
   4. Petioles of cauline leaves ciliate at base ------------- 4. *C. hirsuta*
   4. Petioles of cauline leaves not ciliate at base -------------------- 5
5. Terminal leaflet broader than the lateral leaflets; bases of leaflets decurrent along the rachis ----------------------- 5. *C. pennsylvanica*
5. Terminal leaflet about the same width as the lateral leaflets; bases of leaflets not decurrent along the rachis ----------------------------------
   ---------------------------------6. *C. parviflora* var. *arenicola*

1. **Cardamine bulbosa** (Schreb.) BSP. Prel. Cat. N. Y. 4. 1888.
   Fig. 61.
   Arabis bulbosa Schreb. ex Muhl. Trans. Am. Phil. Soc. 3:174.
   1793.
   Cardamine rhomboidea DC. Syst. Veg. 2:246. 1821.

Perennial from a swollen bulbous base and from tuber-bearing rootstocks; stems erect, usually unbranched, to 50 cm tall, glabrous; basal leaves oblong to nearly orbicular, usually cordate at the base, glabrous, the blade to 3 cm long, the glabrous petiole usually longer than the blade; cauline leaves ovate to lanceolate, subacute at the apex, rounded or subcuneate at the base, entire to dentate, to 4.5 cm long, glabrous, all but the lowest few sessile; racemes terminal, the flowers several, to 1.5 cm broad, on strongly ascending, glabrous pedicels to 2 cm long; sepals 4, elliptic to lanceolate, 2.5–5.0 mm long, green with a white margin; petals 4, white, to 1.5 cm long; siliques nearly straight, erect, linear to narrowly lanceoloid, to 2.5 cm long, the style persistent as a beak 2–4 mm long, tipped by a prominent stigma, the seeds several, oval.

COMMON NAMES: Bulbous Cress; Spring Cress.
HABITAT: Wet soil in woods and along streams.
RANGE: Quebec to Ontario and South Dakota, south to Texas and Florida.
ILLINOIS DISTRIBUTION: Occasional to common throughout the state.
This species and *C. douglassii* are the only species of *Cardamine* in Illinois which lack pinnate or pinnatifid

61. *Cardamine bulbosa* (Bulbous Cress). *a.* Habit, X½. *b.* Flower (with one sepal and one petal removed), X3½. *c.* Sepal, X7½. *d.* Petal, X5. *e.* Fruiting raceme, X¼. *f.* Fruit, X3½. *g.* Seed, X35.

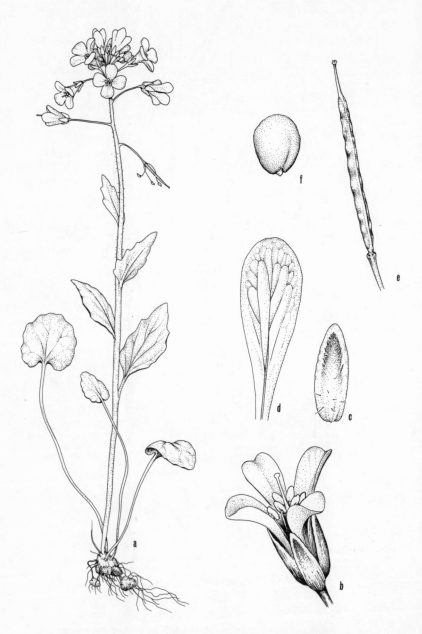

62. *Cardamine douglassii* (Purple Cress). *a.* Habit, X½. *b.* Flower, X5. *c.* Sepal, X7½. *d.* Petal, X4. *e.* Fruit, X3½. *f.* Seed, X35.

leaves. *Cardamine bulbosa* is easily distinguished from *C. douglassii* by its smooth stems, green sepals, and white flowers.

Most parts of the young plant have a horseradish flavor.

The bulbous cress is found in most moist habitats in Illinois. It flowers from late March to June.

2. **Cardamine douglassii** (Torr.) Britt. Trans. N. Y. Acad. Sci. 9:8. 1889. *Fig. 62.*

Arabis rhomboidea Pers. var. *purpurea* Torr. Am. Journ. Sci. 4:66. 1822.

Arabis douglassii Torr. in Torr. & Gray, Fl. N. Am. 1:83. 1838, *pro syn.*

Cardamine rhomboidea (Pers.) DC. var. *purpurea* (Torr.) Torr. Fl. N. Y. 1:56. 1843.

Cardamine bulbosa (Schreb.) BSP. var. *purpurea* (Torr.) BSP. Prel. Cat. N. Y. 4. 1888.

Perennial from a swollen bulbous base and from tuber-bearing root-stocks; stems erect, usually unbranched, to 30 cm long, hirsute, at least in the upper half; basal leaves ovate to orbicular, cordate at the base, glabrous or sparsely pubescent, the blade to 2.5 cm long, angulate-lobed, the petiole longer than the blade; cauline leaves ovate, obtuse to subacute at the apex, rounded to subcordate at the base, entire to dentate, to 3 cm long, glabrous or sparsely pubescent, all but the lowest 1–2 sessile; racemes terminal, the flowers several, to 2 cm broad, on spreading-ascending pedicels to 2.5 cm long; sepals 4, elliptic to lanceolate, 3–6 mm long, purple; petals 4, rose-purple or pink, to 2 cm long; siliques nearly straight, ascending, linear to narrowly lanceoloid, to 2.5 cm long, the style persistent as beak 3–5 mm long, tipped by a prominent stigma, the seeds several, oval.

COMMON NAMES: Purple Cress; Purple Spring Cress; Northern Bitter Cress.

HABITAT: Low woods.

RANGE: Connecticut to Wisconsin, south to Missouri and Virginia; also southern Ontario.

ILLINOIS DISTRIBUTION: Occasional in the northern two-thirds of the state; also Johnson County.

There are enough similarities between *C. douglassii* and *C. bulbosa* that some botanists consider one to be a variety of the other. *Cardamine douglassii* differs by its purplish flowers and hirsute stems.

Where *C. douglassii* and *C. bulbosa* grow in the same vicinity, *C. douglassii* tends to bloom several days earlier. It flowers during April and May.

3. Cardamine pratensis L. var. palustris Wimm. & Grab. Fl. Siles. 1(2):266. 1829. *Fig. 63.*

Perennial herb from a short rhizome; stems ascending to erect, mostly unbranched, to 0.5 m tall, glabrous; leaves pinnately divided into 7–15 leaflets, glabrous, the leaflets of the lower leaves larger, broader, entire to dentate, the leaflets of the upper leaves smaller, narrower, entire, sessile; racemes terminal, the flowers 10–15 mm broad, on ascending pedicels; sepals 4, ovate-lanceolate to elliptic, glabrous, green, 3–5 mm long; petals 4, white, obovate, 8–15 mm long; stamens 6; siliques straight, ascending to erect, linear, to 3 cm long, the style minute, the seeds numerous, wingless.

COMMON NAME: Cuckoo-flower.
HABITAT: Damp areas.
RANGE: Labrador to British Columbia, south to Minnesota, northern Illinois, and western Virginia.
ILLINOIS DISTRIBUTION: Known only from Lake and McHenry counties.

The Illinois stations were discovered by Keith Wilson in 1976 in damp areas of Lake and McHenry counties. Fernald (1950) had attributed this taxon to northern Illinois, but I have been unable to determine the basis for his statement.

This taxon has large flowers similar to the flowers of *C. bulbosa*, but this latter species does not have pinnately divided leaves.

The flowers bloom during June and July.

4. Cardamine hirsuta L. Sp. Pl. 655. 1753. *Fig. 64.*

Annual or biennial herb from fibrous roots; stems ascending to erect, mostly unbranched, to 25 cm tall, glabrous or pubescent near the base; basal leaves several, to 10 cm long, on hirsute petioles, pinnately divided into 3–7 leaflets, the leaflets ovate to orbicular, entire or sparsely toothed, petiolulate, usually hirsutulous; cauline leaves smaller, the 7–11 leaflets linear or linear-oblong, entire, strigose on the upper surface, the petioles ciliate; racemes terminal, the flowers few, 3–5 mm broad, on ascending pedicels; sepals 4, elliptic, hirsutulous near tip, green, 0.5–1.0 mm long; petals 4, white, 1.5–2.0 mm long; stamens 4; siliques straight, strongly as-

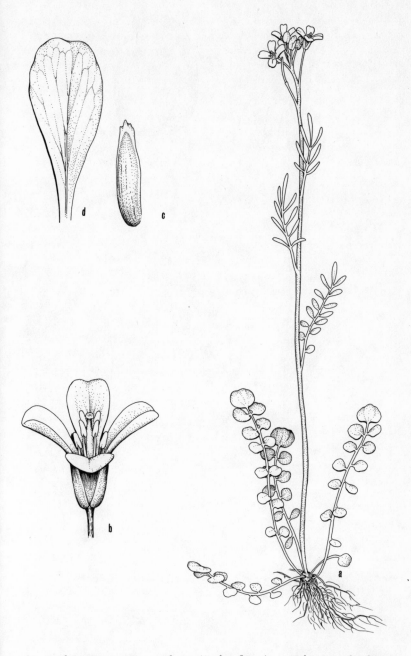

63. Cardamine pratensis var. *palustris* (Cuckoo-flower). *a.* Habit, X½. *b.* Flower, X5. *c.* Sepal, X7½. *d.* Petal, X5.

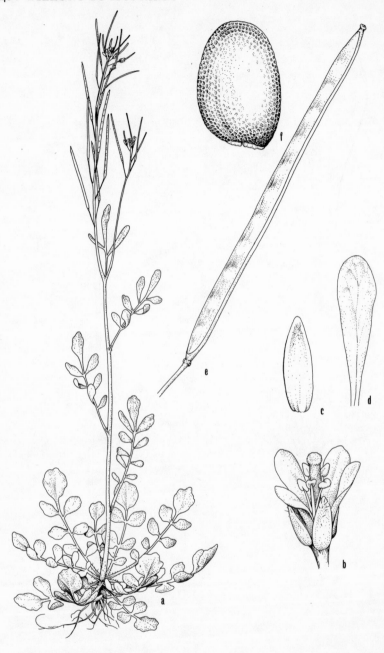

64. *Cardamine hirsuta* (Hairy Bitter Cress). *a*. Habit, X¾. *b*. Flower, X12½.
c. Sepal, X25. *d*. Petal, X20. *e*. Fruit, X5. *f*. Seed, X30.

cending to erect, linear, to 2.5 cm long, the style minute, the seeds 20 or more, wingless.

COMMON NAME: Hairy Bitter Cress.
HABITAT: Wet ground.
RANGE: Native of Europe and Asia; occasionally adventive in the eastern half of the United States.
ILLINOIS DISTRIBUTION: Scattered in most parts of Illinois. The hairy bitter cress is confused more often with *Sibara virginica* (L.) Rollins than with other species of *Cardamine*. Both species usually have some hispidity at the base of the stem and erect, linear siliques. The seeds of *C. hirsuta* are wingless, while those of *Sibara virginica* are winged.

Cardamine hirsuta is primarily a species of wet, disturbed areas. It flowers during March and April.

5. **Cardamine pennsylvanica** Muhl. ex. Willd. Sp. Pl. 3:486. 1800. *Fig. 65.*

Annual or biennial herb from fibrous roots; stems erect, usually branched, to 75 cm tall, glabrous; basal leaves several, to 15 cm long, on glabrous petioles, pinnately divided into 3–17 leaflets, the terminal leaflet obovate to suborbicular, larger and broader than the lateral leaflets, dentate or undulate, rarely entire, glabrous; cauline leaves pinnately divided into 9–17 leaflets, the leaflets linear to oblanceolate, entire or dentate or undulate, glabrous; racemes terminal, several-flowered, the flowers 4–5 mm broad, on ascending pedicels; sepals 4, elliptic, glabrous, 1–2 mm long, green; petals 4, white, 2–4 mm long; stamens 6; siliques straight, erect, linear, glabrous, to 3.5 cm long, the slender style persistent as a beak up to 2 mm long, the seeds numerous, 1.0–1.5 mm long, wingless.

COMMON NAME: Bitter Cress.
HABITAT: Low woods; base of wet cliffs; damp fields.
RANGE: Labrador to British Columbia, south to Oregon, Texas, and Florida.
ILLINOIS DISTRIBUTION: Scattered in most sections of the state.
Cardamine pennsylvanica is a variable species in the shape of the terminal leaflets and in their texture. Plants growing in deep shade generally have larger, membranous leaflets.

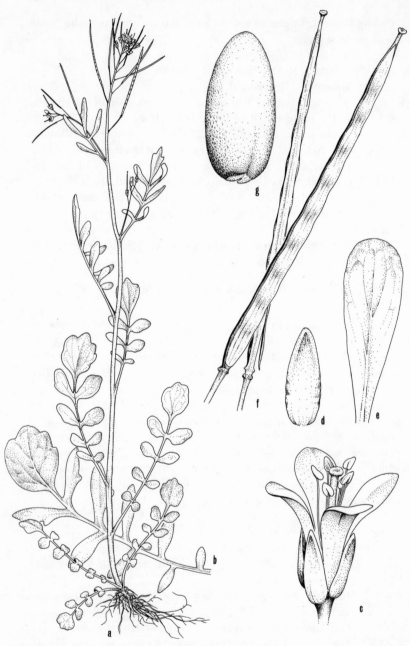

65. *Cardamine pennsylvanica* (Bitter Cress). *a.* Habit, X¾. *b.* Leaf, X1½. *c.* Flower, X10. *d.* Sepal, X15. *e.* Petal, X12½. *f.* Fruits, X4. *g.* Seed, X35.

This species differs from *C. hirsuta* by its glabrous stems, and from *C. parviflora* var. *arenicola* by its broader terminal leaflets.

Although Fernald (1950) and others attribute hispid-pubescence to the base of the stem in *C. pennsylvanica*, stems in Illinois material, at least, lack it.

Almost all Illinois botanists during the nineteenth century erroneously called this species *C. hirsuta*.

Fresh stems and leaves may be used in salads.

The flowering period for *C. pennsylvanica* is March to July.

6. **Cardamine parviflora** L. var. **arenicola** (Britt.) O. E. Schulz, Bot. Jahrb. 32:485. 1903. *Fig. 66.*

Cardamine virginica Michx. Fl. Bor. Am. 2:29. 1803, non L. (1753).

Cardamine arenicola Britt. Bull. Torrey Club 19:220. 1892.

Annual or biennial herbs from fibrous roots; stems erect, usually branched, to 35 cm tall, glabrous; basal leaves several, to 10 cm long, on glabrous petioles, pinnately divided into 5–13 leaflets, the leaflets all nearly uniform in size and shape, linear to linear-oblong or occasionally obovate, entire or few-toothed, glabrous; cauline leaves pinnately divided into 5–13 usually uniform leaflets, linear to oblanceolate, entire, glabrous; racemes terminal, several-flowered, the flowers 4–5 mm broad, on ascending pedicels; sepals 4, elliptic, glabrous, 1–2 mm long, green; petals 4, 2–4 mm long; stamens 6; siliques straight, erect, linear, glabrous, to 3 cm long, the short slender style persistent as a beak up to 1 mm long, the seeds numerous, 1.0–1.5 mm long, wingless.

COMMON NAME: Small-flowered Bitter Cress.

HABITAT: Dry woods and fields; wet ledges or depressions in cliffs.

RANGE: Nova Scotia to Ontario, south to Texas and Florida; British Columbia to Oregon.

ILLINOIS DISTRIBUTION: Scattered throughout the state.

Our plants constitute a variety of the European *C. parviflora* L. Some botanists consider our plants to be the same as *C. parviflora*. At the opposite extreme, some botanists give them separate specific status, calling them *C. arenicola*.

Whatever the name for our plant, it can be differentiated from *C. pennsylvanica* by the uniform leaflets. In *C. pennsylvanica*, the terminal leaflet is larger than the lateral ones.

66. *Cardamine parviflora* var. *arenicola* (Small-flowered Bitter Cress). *a.* Habit,
X¾. *b.* Habit (in silhouette), X¼. *c.* Leaf, X1½. *d.* Flower, X10. *e.* Sepal, X15.
f. Petal, X12½. *g.* Fruit, X5. *h.* Seed, X30.

This taxon usually occupies somewhat drier sites than *C. pennsylvanica*, although it does occasionally grow in wet depressions and ledges of sandstone cliffs.

Fresh stems and leaves of this plant may also be used in salads. *Cardamine parviflora* var. *arenicola* flowers from March to July.

11. *Sibara* Greene – Rock Cress

Annual or biennial herbs from fibrous roots; leaves cauline and basal, pinnately divided; inflorescence racemose, with white flowers; flowers actinomorphic, mostly bractless; sepals 4, free; petals 4, free; stamens 6; pistil one, the ovary superior, the style small; siliques linear, flat, the valves faintly 1-nerved or nerveless, with several winged seeds in one row in each cell.

Sibara is seemingly intermediate between *Cardamine* and *Arabis*. It has the pinnatifid leaves like most species of *Cardamine*, but winged seeds like most species of *Arabis*. Greene derived the generic name by spelling *Arabis* backward.

Only the following species occurs in Illinois.

1. **Sibara virginica** (L.) Rollins, Rhodora 43:481. 1941. *Fig. 67.*
Cardamine virginica L. Sp. Pl. 656. 1753.
Arabis virginica (L.) Poir. in Lam. Encycl. 1:413. 1810.
Cardamine ludoviciana Hook. Journ. Bot. 1:191. 1834.
Arabis ludoviciana (Hook.) Meyer, Ind. Sem. Petr. 9:60. 1842.

Annual or biennial herb from fibrous roots; stems spreading to ascending, stiffly branched, to 30 cm long, hispid at the base, otherwise glabrous; basal leaves several, to 8 cm long, on hispidulous petioles, lyrate-pinnatifid into as many as 21 divisions, the segments linear to oblong, undulate or entire, glabrous or nearly so; cauline leaves similar; racemes terminal, several-flowered, the flowers 2–4 mm broad, on stiffly ascending pedicels to 3 mm long; sepals 4, narrowly elliptic, glabrous, green, 1–2 mm long; petals 4, white, 2–3 mm long; siliques straight or slightly arcuate, linear, flat, glabrous, to 2.5 cm long, 1–2 mm broad, on pedicels up to 4 mm long, the valves faintly 1-nerved, the style persistent as a very short beak, the seeds numerous, winged, orbicular, about 1.5 mm in diameter, arranged in one row in the cell.

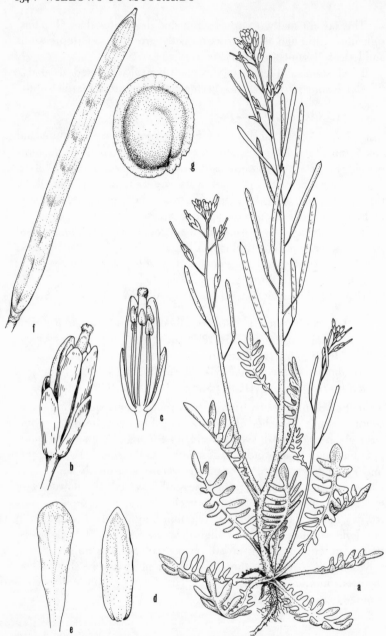

67. *Sibara virginica* (Virginia Rock Cress). *a.* Habit, X¾. *b.* Flower, X7½. *c.*
Flower (with one sepal and two petals removed), X7½. *d.* Sepal, X20. *e.* Petal,
X12½. *f.* Fruit, X5. *g.* Seed, X15.

COMMON NAME: Virginia Rock Cress.

HABITAT: Open woods, fallow fields, waste ground.

RANGE: Virginia across southeastern Kansas to California, south to Texas and Florida; Mexico.

ILLINOIS DISTRIBUTION: Occasional in the southern three-fifths of Illinois; Cook County.

This species, with its pinnately divided leaves, has the appearance of *C. parviflora* var. *arenicola* and *C. pennsylvanica*, except for the hispid base of its stems. From *Cardamine hirsuta* it is distinguished by its winged seeds.

Sibara virginica occupies disturbed areas and usually is associated with common weedy species.

For those not recognizing *Sibara* as a distinct genus, *Arabis virginica* (L.) Poir. is the binomial used for this species.

Arabis ludoviciana (Hook.) Meyer, by which this species was known during the nineteenth century by Illinois botanists, does not have priority as the proper binomial.

The Virginia rock cress flowers from March to May.

12. *Draba* L.—Whitlow Grass

Annual (in Illinois), biennial, or perennial herbs, often with stellate pubescence; leaves basal or cauline and alternate or both, entire or toothed; inflorescence racemose, with white (in Illinois) or yellow flowers; flowers actinomorphic, bractless; sepals 4, free; petals 4, free; stamens 6; pistil one, the ovary superior, the style short; fruits either siliques or silicles, flat, the valves nerveless, with several wingless seeds arranged in two rows in each cell.

Over 150 species of *Draba* occur in the world, primarily in north temperate and Arctic regions. A few species occur in southern South America.

Draba is the only genus of Brassicaceae in the eastern United States with the combination of stellate pubescence and short fruits with several seeds.

KEY TO THE SPECIES OF Draba IN ILLINOIS

1. No leaves present on flowering scape; petals deeply 2-cleft _____1. *D. verna*
1. One or more leaves present on flowering stem _____ 2
  2. Leaves confined to lower part of stem _____ 3
  2. Leaves all along stem _____ 4
3. Leaves dentate; pedicels pubescent _____ 2. *D. cuneifolia*
3. Leaves entire; pedicels glabrous _____ 3. *D. reptans*

4. Fruits to 5 mm long, glabrous, with up to 15 seeds _____
_____4. *D. brachycarpa*
4. Fruits at least 6 mm long, pubescent, with 20 or more seeds _____
_____2. *D. cuneifolia*

1. Draba verna L. Sp. Pl. 642. 1753.

Annual from a small tuft of fibrous roots; leaves crowded in a basal cluster, oblanceolate to oblong, subacute to acute at the apex, cuneate at the base, entire or sparsely dentate, to 2.5 cm long, both surfaces covered with unbranched as well as stellate hairs; scapes 1-several from the base, erect to ascending, to 15 cm tall, glabrous or pubescent near base, bearing an elongated raceme of flowers, each flower 3.0–4.5 mm broad, bractless, on ascending pedicels; sepals 4, broadly elliptic, pubescent, green, 1–2 mm long; petals 4, bifid for half their length, white, 1.5–2.5 mm long; silicles oblong-ellipsoid to obovoid, glabrous, 2.5–10.0 mm long, 1.5–4.0 mm broad, shorter than the pedicels, tipped by the minute persistent style, with up to 60 seeds.

Two varieties occur in Illinois.

1. Seeds 40 or more; fruits more than twice as long as broad _____
_____1a. *D. verna* var. *verna*
1. Seeds less than 40; fruits never more than twice as long as broad _____
_____ 1b. *D. verna* var. *boerhaavii*

1a. Draba verna L. var. **verna** *Fig. 68a–f.*

Seeds 40 or more; silicles more than twice as long as broad, up to 10 mm long, to 2.5 mm broad.

COMMON NAMES: Vernal Whitlow Grass; Mouse-eared Whitlow Grass.
HABITAT: Lawns, pastures, fields, roadsides.
RANGE: Native of Europe and Asia; naturalized in the eastern half of the United States.
ILLINOIS DISTRIBUTION: Occasional and scattered in the state, except for the northwestern counties.
This tiny whitlow grass is readily distinguished by its deeply bifid petals. I have observed flowering specimens as small as 1 cm tall.

Draba verna var. *verna* occurs in a wide range of open waste ground. It has increased greatly in abundance since about 1955.

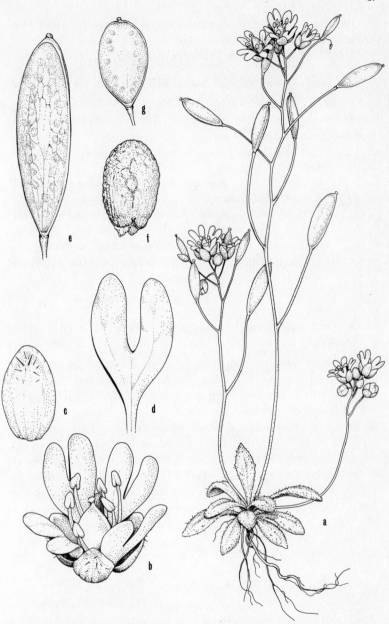

68. *Draba verna* (Vernal Whitlow Grass). *a.* Habit, X2. *b.* Flower, X10. *c.* Sepal, X15. *d.* Petal, X15. *e.* Fruit, X5. *f.* Seed, X25. var. *boerhaavii.* *g.* Fruit, X5.

Despite its small stature, this plant has been used to treat whitlow disease in animals' hooves.

The flowers bloom from February to April.

1b. Draba verna L. var. **boerhaavii** Van Hall, Specim. Bot. 149. 1821. *Fig. 68g.*

Seeds less than 40; silicles never more than twice as long as broad, to 6 mm long, to 4 mm broad.

HABITAT: Lawns.

RANGE: Native of Europe; naturalized in the eastern half of the United States.

ILLINOIS DISTRIBUTION: Known only from Jackson County.

This variety is much less common than var. *verna*, although it is probably more common than the map indicates. It has very short silicles.

2. Draba cuneifolia Nutt. ex Torr. & Gray, Fl. N. Am. 1:108. 1838.

Annual from a tuft of fibrous roots; stems branched from near the base, hispid with unbranched and stellate hairs, to 20 cm tall; leaves crowded in a basal cluster and at the lower nodes, rarely along the entire length of the stem, obovate or spatulate, obtuse to subacute at the apex, cuneate at the base, dentate, to 3 cm long, pubescent on both surfaces with unbranched and stellate hairs; racemes terminal, at length loosely flowered, the flowers 3.5–4.5 mm broad, on spreading to ascending pedicels; sepals 4, elliptic, pubescent, green, 1.0–1.5 mm long; petals 4, white, some of them 3–5 mm long, others up to 2 mm long, sometimes absent; silicles linear to oblongoid, pubescent, flat, up to 1.5 cm long, the style absent, the spreading, pubescent pedicels to 1 cm long, with 20 or more seeds.

Two varieties occur in Illinois.

1. Leaves basal or only from the lowest nodes on the stems _____
_____2a. *D. cuneifolia* var. *cuneifolia*
1. Leaves cauline as well as basal _____ 2b. *D. cuneifolia* var. *foliosa*

2a. Draba cuneifolia Nutt. var. **cuneifolia** *Fig. 69a–f.*

Leaves basal and sometimes with leaves from the lowest nodes.

COMMON NAME: Whitlow Grass.

HABITAT: Limestone ledges.

RANGE: Kentucky and southern Illinois to Colorado, south to Texas and Florida; Mexico.

ILLINOIS DISTRIBUTION: Known only from the western edge of Jersey, Madison, Monroe, Randolph, and St. Clair counties.

This whitlow grass resembles *D. reptans* by its leaves nearly all clustered at the base of the plant and the lower part of the stem, but differs by its entire leaves and glabrous pedicels.

Draba cuneifolia is restricted to limestone ledges at the edge of hill prairies between Prairie du Rocher (Randolph County) and Fults (Monroe County), and near Elsah (Madison County).

The flowers open from February to May. Most of the flowers have petals 3–5 mm long, but sometimes the flowers are only 1–2 mm long. In even rarer specimens, the petals may be absent.

Higley and Raddin (1891) reported this species from Chicago, based on a Babcock collection made in April, 1875, probably a waif. I have not been able to locate this specimen. The first surely native specimen collected was from a cliff near Elsah, Madison County, on May 18, 1918, by R. Hoffmann.

2b. Draba cuneifolia Nutt. var. **foliosa** Mohlenbrock, Rhodora 62:240. 1960. *Fig. 69g.*

Leaves on the upper part of the stem as well as the lower.

COMMON NAME: Wedge-leaved Whitlow Grass.

HABITAT: Limestone ledge.

RANGE: Southwestern Illinois.

ILLINOIS DISTRIBUTION: Known only from its original collection in Randolph County (edge of limestone bluff, one mile northwest of Prairie du Rocher, T5S, R9E, section 16, *R. H. Mohlenbrock 5969*).

This unusual variety grows on the same ledge as does *D. cuneifolia* var. *cuneifolia* and *D. reptans*. The plants look more like *D. aprica* Beadle, but the silicles are over 6 mm long and contain more than 20 seeds.

3. Draba reptans (Lam.) Fern. Rhodora 36:368. 1934.

Arabis reptans Lam. Encycl. 1:222. 1783.

Annual from a tuft of fibrous roots; stems branched from near the

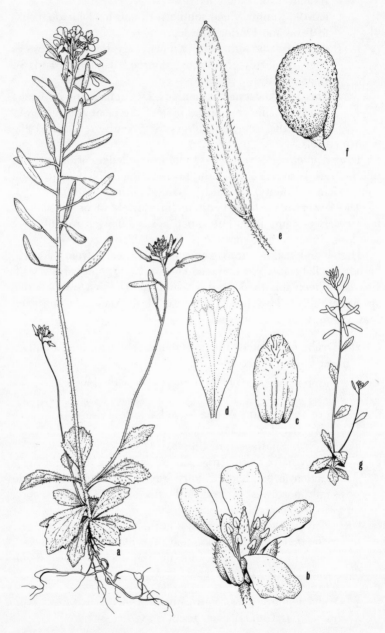

69. *Draba cuneifolia* (Whitlow Grass). *a.* Habit, X1. *b.* Flower, X10. *c.* Sepal, X20. *d.* Petal, X10. *e.* Fruit, X5. *f.* Seed, X30. var. *foliosa.* *g.* Habit, X¼.

base, erect to ascending, hispid below with unbranched and stellate
hairs, glabrous above, to 15 cm tall; leaves crowded in a basal cluster and at the lower nodes, obovate to spatulate, obtuse at the apex,
cuneate or somewhat rounded at the base, entire or less commonly
sparsely dentate, to 2 cm long, bristly ciliate, stellate-pubescent, at
least on the lower surface; racemes terminal, the several flowers
clustered near the tip of the stem, the flowers 3.0–4.5 mm broad,
on glabrous, ascending pedicels; sepals 4, elliptic, glabrous or nearly
so, green, 1–2 mm long; petals 4, white, 2–5 mm long, or absent;
silicles straight, linear to narrowly oblongoid, glabrous or hispidulous, flat, up to 2 cm long, the style absent, the ascending pedicels
to 5 mm long, with 15 or more seeds.

Two varieties occur in Illinois.

1. Silicles glabrous _____ 3a. *D. reptans* var. *reptans*
1. Silicles hispidulous _____ 3b. *D. reptans* var. *micrantha*

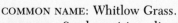

3a. Draba reptans (Lam.) Fern. var. **reptans** *Fig. 70.*
Draba caroliniana Walt. Fl. Car. 174. 1788.

Silicles glabrous.

COMMON NAME: Whitlow Grass.

HABITAT: Sand prairies; limestone ledges; limestone
glades; rarely roadsides.

RANGE: Massachusetts to North Dakota, south to Colorado, New Mexico, and Georgia; Washington and Oregon.

ILLINOIS DISTRIBUTION: Occasional in the northern counties; less common in the southern counties.

This tiny whitlow grass differs from *D. cuneifolia* by its
nearly entire leaves and glabrous pedicels and silicles.

In the northern half of the state, this variety is found in
sandy prairies, hill prairies, and limestone glades. Swink has shown
it to me in disturbed roadsides, growing with *Androsace occiden-
talis*. In southwestern Illinois, it grows on limestone ledges with
D. cuneifolia.

Until 1934, this plant was known as *D. caroliniana* Walt.

The flowers, which sometimes are apetalous, bloom from February to May.

3b. Draba reptans (Lam.) Fern. var. **micrantha** (Nutt.) Fern.
Rhodora 36:368. 1934.
Draba micrantha Nutt. ex Torr. & Gray, Fl. N. Am. 1:109.
1838.

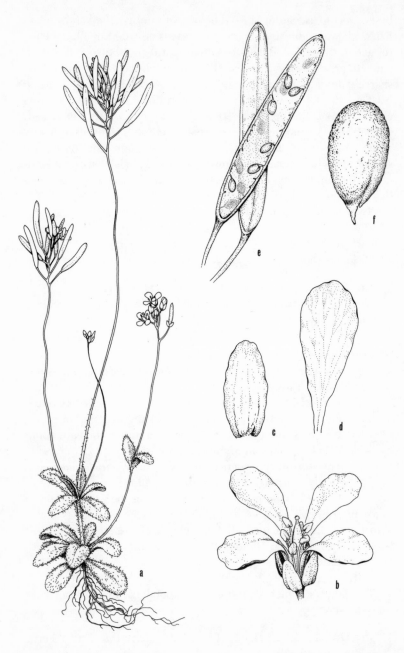

70. *Draba reptans* (Whitlow Grass). *a*. Habit, X1. *b*. Flower, X10. *c*. Sepal, X20. *d*. Petal, X10. *e*. Fruits, X3½. *f*. Seed, X30.

Draba caroliniana Walt. var. *micrantha* (Nutt.) Gray, Man. Bot. ed. 5, 72. 1867.

Silicles hispidulous.

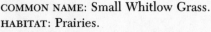

COMMON NAME: Small Whitlow Grass.

HABITAT: Prairies.

RANGE: Minnesota to Washington, south to California, Colorado, and northern Illinois.

ILLINOIS DISTRIBUTION: Occasional in a few northern counties.

This variety is much less common than var. *reptans*.

4. Draba brachycarpa Nutt. ex Torr. & Gray, Fl. N. Am. 1:108. 1838. *Fig. 71*.

Tufted annual from a cluster of fibrous roots; stems simple or branched from near the base, erect to ascending, stellate-pubescent, to 15 cm tall; basal leaves ovate to obovate, obtuse at the apex, cuneate to the short-petiolate base, entire to sparsely dentate, to 2 cm long, stellate-pubescent; cauline leaves elliptic to oblong, obtuse at the apex, cuneate to the sessile base, to 1.5 cm long, stellate-pubescent; racemes terminal or from the uppermost axils, bracteate, with several crowded, bractless flowers, the flowers 2–3 mm broad, on spreading to ascending pedicels; sepals 4, narrowly elliptic, glabrous or nearly so, green, 0.5–2.0 mm long; petals 4, white, 2–3 mm long, or absent; silicles oblongoid to ellipsoid, acute at the tip, flat, glabrous, to 5 mm long, to 1.5 mm broad, the style minute, the spreading to ascending pedicels to 5 mm long, with 10–15 seeds.

COMMON NAME: Short-fruited Whitlow Grass.

HABITAT: Fields, lawns, prairies, woods, along railroads.

RANGE: Virginia to Kansas, south to Texas and Florida; Oregon.

ILLINOIS DISTRIBUTION: Common in the southern two-fifths of the state; rare elsewhere.

Although this species grows in weedy habitats, it is considered to be native in southern Illinois. The DeKalb County collection probably was made from an adventive plant.

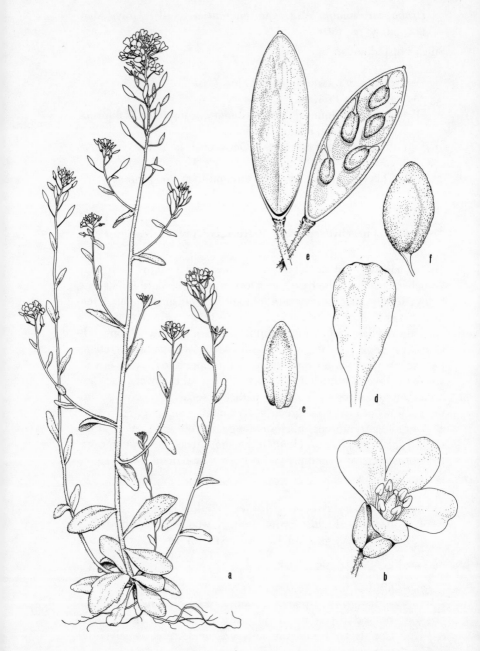

71. *Draba brachycarpa* (Short-fruited Whitlow Grass). *a.* Habit, X1. *b.* Flower, X15. *c.* Sepal, X20. *d.* Petal, X15. *e.* Fruits, X10. *f.* Seed, X20.

Many of the flowers are apetalous. The flowers bloom from March to May.

13. *Arabidopsis Heynh.* – Mouse-eared Cress

Annual or perennial herbs, usually with forked hairs; leaves basal and cauline, entire or toothed; inflorescence racemose, with white (in Illinois), yellow, or pink flowers; flowers actinomorphic, bractless; sepals 4, free; petals 4, free; stamens 6; pistil one, the ovary superior, the style short; siliques narrowly linear, cylindrical, the valves reticulate-nerved, with several winged seeds arranged in 1–2 rows in each cell.

There are about one dozen species of *Arabidopsis* native to Europe and Asia. The genus differs from *Arabis* in technical characters of the cotyledons.

Only the following species occurs in Illinois.

1. Arabidopsis thaliana (L.) Heynh. in Holl & Heynh. Fl. Sachs. 1:538. 1842. *Fig. 72.*
Arabis thaliana L. Sp. Pl. 665. 1753.

Annual from a slender, elongated root; stems simple or branched, erect, pubescent with unbranched and forked hairs, to 45 cm tall; basal leaves clustered, oblanceolate to oblong, obtuse at the apex, cuneate to the petiolate base, entire or toothed, pubescent, to 4 cm long; cauline leaves alternate, remote, elliptic, subacute to acute at the apex, cuneate to the sessile base, entire or sparsely toothed, pubescent, to 1.5 cm long; racemes terminal and from the axils of the upper leaves, the several flowers densely crowded at first, each flower 2.5–4.0 mm broad, on nearly filiform, spreading to ascending pedicels; sepals 4, broadly elliptic, glabrous or nearly so, green, 1.5–3.0 mm long; petals 4, white, usually emarginate, spatulate, 3–4 mm long; siliques straight or arcuate, cylindrical, narrowly linear, glabrous, to 1.5 cm long, less than 1 mm wide, the style absent, the loosely ascending pedicels to 1.5 cm long, with 40 or more minute seeds arranged in one row.

COMMON NAME: Mouse-eared Cress.

HABITAT: Fields, pastures, along roads.

RANGE: Native of Europe and Asia; naturalized in the eastern half of the United States.

ILLINOIS DISTRIBUTION: Occasional in the southern half of the state, rare elsewhere.

Mouse-eared cress resembles the genus *Arabis*, but has

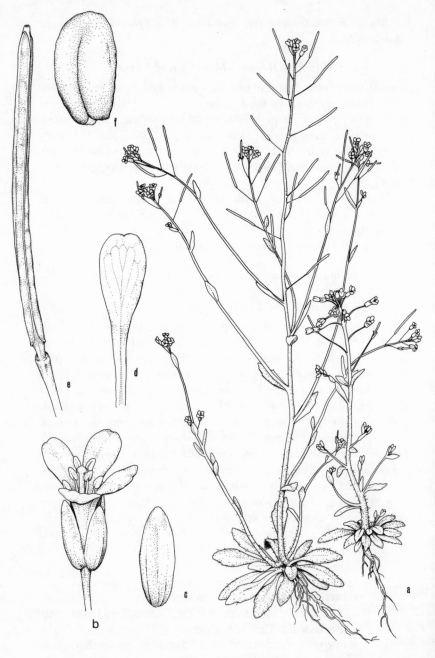

72. *Arabidopsis thaliana* (Mouse-ear Cress).   *a.* Habit, X1.   *b.* Flower, X10.   *c.* Sepal, X12½.   *d.* Petal, X12½.   *e.* Fruit, X6.   *f.* Seed, X37½.

narrower siliques and narrower basal leaves.
It flowers from March to May.

14. Lepidium L.–Peppergrass

Annual, biennial, or perennial herbs, usually without stellate pubescence; leaves entire to toothed to pinnatifid; inflorescence racemose, with white or greenish flowers; flowers actinomorphic, bractless; sepals 4, free; petals 4, free, or absent; stamens 2, 4, or mostly 6; pistil one, the ovary superior, the style short; silicles flattened contrary to the partition, usually emarginate, the valves keeled, with one flattened, pendulous seed per cell.

About 65 species distributed throughout much of the world comprise the genus *Lepidium*. Of the six species in Illinois, all but one is naturalized from Europe and Asia.

KEY TO THE SPECIES OF Lepidium IN ILLINOIS

1. Plants densely hairy _____ 1. *L. campestre*
1. Plants glabrous or puberulent _____ 2
  2. Cauline leaves auriculate at base _____ 2. *L. perfoliatum*
  2. Cauline leaves tapering to base _____ 3
3. Fruits winged their entire length, 5–6 mm long; stamens 6 _____
  _____ 3. *L. sativum*
3. Fruits unwinged, or winged only at tip, 2–4 mm long; stamens 2 (–4)
  _____4
  4. Petals about as long as the sepals _____ 4. *L. virginicum*
  4. Petals absent _____ 5
5. Plants with a fetid odor; basal leaves bipinnatifid _____ 5. *L. ruderale*
5. Plants without a fetid odor; basal leaves once-pinnatifid or coarsely
  toothed _____6. *L. densiflorum*

1. **Lepidium campestre** (L.) R. Br. in Ait. f. Hort. Kew. 4:88.
  1812. *Fig. 73.*
  Thlaspi campestre L. Sp. Pl. 646. 1753.

Annual or biennial herb from a somewhat thickened root; stems erect, sometimes branched above, to 50 cm tall, densely spreading-pubescent, gray-green; basal leaves spatulate to oblong, obtuse at the apex, cuneate to the petiolate base, entire or the lower part of the leaf pinnatifid, to 7 cm long, hoary-pubescent, gray-green; cauline leaves lanceolate to oblong, acute to acuminate at the apex, sagittate at the clasping base, denticulate, to 4 cm long, hoary-pubescent; racemes terminal, densely flowered, the flowers 2.5–4.0 mm broad, on stout, spreading-ascending pedicels; sepals 4, broadly

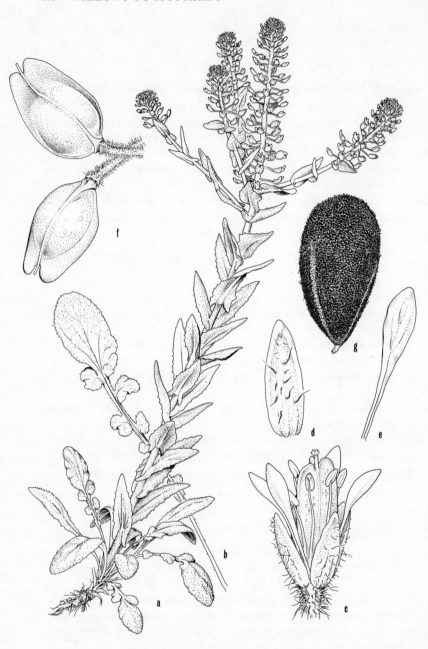

73. *Lepidium campestre* (Field Peppergrass). *a*. Habit, X½. *b*. Basal leaf, X1½. *c*. Flower, 12½. *d*. Sepal, X20. *e*. Petal, X15. *f*. Fruits, X5. *g*. Seed, X25.

lanceolate, glabrous or nearly so, green, 1–2 mm long; petals 4, white, 2–3 mm long; stamens 6; silicles oblong-ovoid, emarginate, 5–6 mm long, glabrous, sometimes papillose, broadly winged at the apex, the style persistent in the notch, the pedicels spreading, 4–8 mm long.

COMMON NAMES: Field Peppergrass; Field Cress; Field Peppercress; Cow Cress.

HABITAT: Roadsides, fields, pastures, and along railroads.

RANGE: Native of Europe; naturalized in much of North America.

ILLINOIS DISTRIBUTION: Occasional and scattered throughout the state.

The dense, spreading pubescence of the stems and leaves readily distinguishes this species from others of the genus in Illinois. The cauline leaves strongly overlap on the stem.

This species is more abundant in the northern half of the state than in the southern half.

The flowers open from April to June.

2. Lepidium perfoliatum L. Sp. Pl. 643. 1753. *Fig. 74.*

Annual herb from a slightly thickened, elongated root; stems erect, usually branched, to 50 cm tall, glabrous; basal leaves bipinnate to bipinnatifid, the ultimate segments linear, glabrous; cauline leaves ovate, acute at the apex, cordate at the clasping or perfoliate base, entire, glabrous, to 3 cm long; racemes terminal, densely flowered, the flowers 2.0–3.5 mm broad, on glabrous, spreading-ascending pedicels; sepals 4, broadly lanceolate, glabrous, green, 1.5–2.5 mm long; petals 4, white, linear, 2–3 mm long; stamens 6; silicles rhombic-ovoid, subacute to emarginate at the apex, 3–4 mm long, glabrous, the persistent style minute, the pedicels spreading-ascending, 4–10 mm long.

COMMON NAMES: Perfoliate Peppergrass; Clasping Cress.

HABITAT: Along railroads.

RANGE: Native of Europe; naturalized in much of the United States.

ILLINOIS DISTRIBUTION: Scattered in a few northern counties, rare elsewhere.

The bipinnate or bipinnatifid basal leaves contrast remarkably with the simple, entire, clasping or perfoliate cauline leaves.

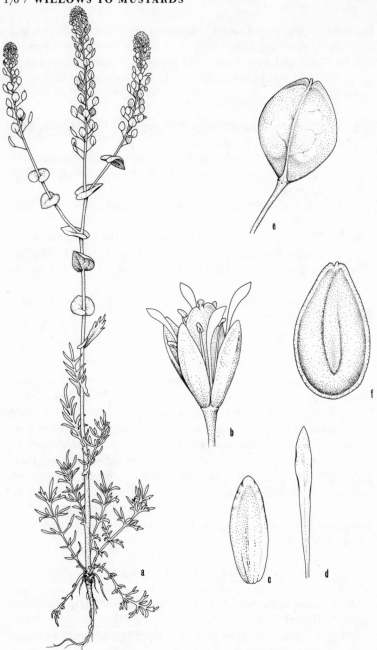

74. *Lepidium perfoliatum* (Perfoliate Peppergrass). *a*. Habit, X½. *b*. Flower, X10. *c*. Sepal, X15. *d*. Petal, X15. *e*. Fruit, X7½. *f*. Seed, X20.

The first Illinois collection apparently was made in 1945.
Swink (1974) reports it to be on the increase, and cites a locality
in Grundy County where thousands of plants occur.
The clasping cress flowers from April to June.

3. **Lepidium sativum** L. Sp. Pl. 644. 1753. *Fig. 75.*

Annual herb from a slightly thickened, elongated root; stems erect,
branched, to 75 cm tall, glabrous, glaucous; basal leaves bipinnatifid, to 15 cm long, long-petiolate, the segments entire or toothed,
glabrous; cauline leaves similar but smaller, sessile or nearly so; racemes terminal, densely flowered, the flowers 1.5–2.0 mm broad,
on glabrous, strongly ascending pedicels; sepals 4, broadly lanceolate, glabrous, green, 1.0–1.5 mm long; petals 4, white, 1.5–2.0
mm long; stamens 6; silicle oblong-ovoid, emarginate, glabrous,
5–6 mm long, winged all the way around, the persistent style in the
notch, the pedicels ascending, as long as or longer than the silicles.

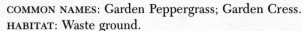

COMMON NAMES: Garden Peppergrass; Garden Cress.
HABITAT: Waste ground.
RANGE: Native of Europe; rarely escaped from cultivation
in North America.
ILLINOIS DISTRIBUTION: Cook Co.: Englewood, July 8,
1875, *E. J. Hill s.n.*; Naperville, July 13, 1897, *L. M.
Umbach s.n.*; also Kendall County.
This species, because of its pungent foliage, is sometimes
cultivated in gardens as a salad plant. It rarely escapes
from cultivation.
This is the only nonauriculate species of *Lepidium* in Illinois
with a wing all the way around the silicle. The pedicels are very
strongly ascending.
The garden peppergrass flowers in May and June.

4. **Lepidium virginicum** L. Sp. Pl. 645. 1753. *Fig. 76.*

Annual or biennial herb from an elongated root; stems erect,
branched or unbranched, to 75 cm tall, glabrous or puberulent;
basal leaves obovate to spatulate, usually pinnatifid with the terminal lobe the largest, glabrous or puberulent; cauline leaves linear
to lanceolate, acute at the apex, cuneate at the usually sessile base,
entire to sharply serrate, glabrous or nearly so; racemes terminal,
sometimes branched, densely flowered, the flowers 1–2 mm broad,
on ascending pedicels; sepals 4, lanceolate, glabrous, green, 1.0–
1.5 mm long; petals 4, white, 1.5–2.5 mm long; stamens 2 (rarely

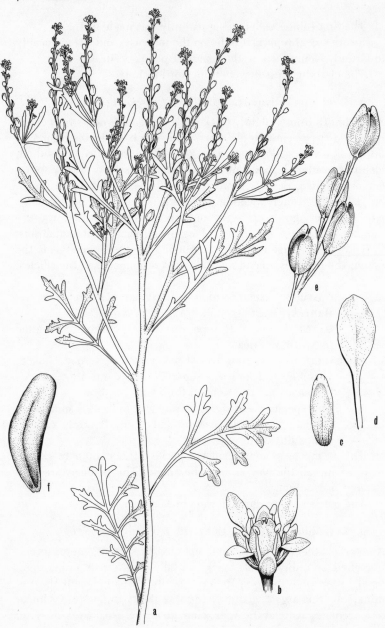

75. *Lepidium sativum* (Garden Peppergrass). *a.* Upper part of plant, X½. *b.* Flower, X12½. *c.* Sepal, X17½. *d.* Petal, X17½. *e.* Cluster of fruits, X2½. *f.* Seed, X20.

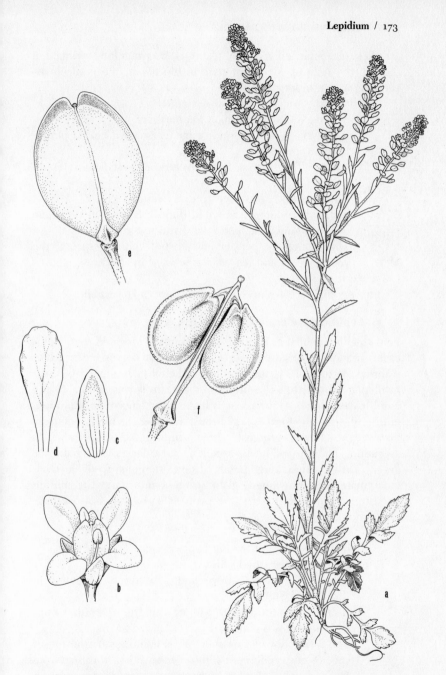

76. *Lepidium virginicum* (Common Peppergrass). *a*. Habit, X½. *b*. Flower, X10.
c. Sepal, X20. *d*. Petal, X15. *e*. Fruit, X12½. *f*. Fruit with valves removed,
exposing two seeds, X15.

4); silicle orbicular, emarginate, glabrous, 2–4 mm long, winged at the tip, the minute persistent style in the notch, the pedicels ascending, 4–7 mm long.

COMMON NAME: Common Peppergrass.
HABITAT: Waste ground, edge of woods, fields, prairie borders, pastures.
RANGE: Throughout North America; West Indies; South America.
ILLINOIS DISTRIBUTION: In every county.
Although reputedly native in Illinois, *L. virginicum* is one of the commonest species of disturbed ground.
The silicles are peppery to the taste. In the spring, the young stems and leaves can be used in salads as a substitute for water cress.

Flowers may be produced from February to December.

5. Lepidium ruderale L. Sp. Pl. 645. 1753. *Fig. 77.*

Foul-smelling annual or biennial herb from an elongated root; stems erect, branched, to 50 cm tall, glabrous; basal leaves bipinnatifid, glabrous, to 10 cm long, long-petiolate, the ultimate segments linear to oblong, obtuse to subacute; lower cauline leaves similar to the basal leaves; upper cauline leaves linear to oblanceolate, unlobed, entire to sparsely toothed, glabrous; racemes terminal, densely flowered, the flowers greenish, about 1 mm broad, on spreading to ascending, glabrous pedicels; sepals 4, linear-lanceolate, glabrous, green, 1.0–1.5 mm long; petals absent; stamens 2; silicle oval, emarginate, flat, unwinged, glabrous, 2–3 mm long, the minute persistent style in the notch, the pedicels spreading to ascending, to 5 mm long.

COMMON NAME: Stinking Peppergrass.
HABITAT: Disturbed soils.
RANGE: Native of Europe and Asia; rarely adventive in the United States.
ILLINOIS DISTRIBUTION: Known only from DeKalb County, where it was collected by Mr. Floyd Swink.
The fetid odor of this species distinguishes it from other species of *Lepidium* in Illinois. In addition, it differs from the somewhat similar *L. densiflorum* by its bipinnatifid basal leaves.

The stinking peppergrass flowers in May and June.

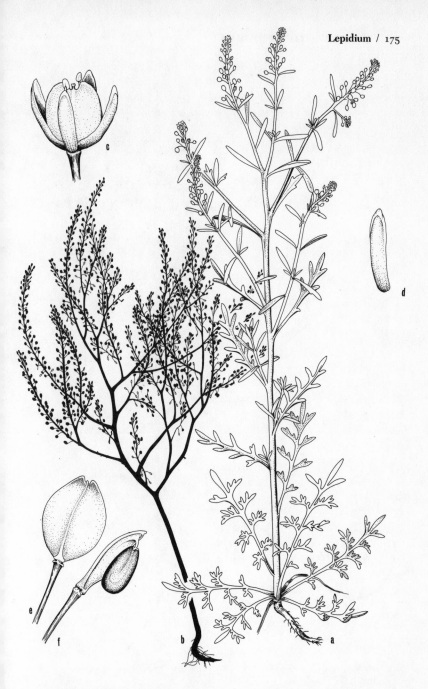

77. *Lepidium ruderale* (Stinking Peppergrass). *a.* Habit, X½. *b.* Habit, in silhouette, X⅓. *c.* Flower, X15. *d.* Sepal, X22½. *e.* Fruit, X10. *f.* Fruit with valves removed, showing one seed, X10.

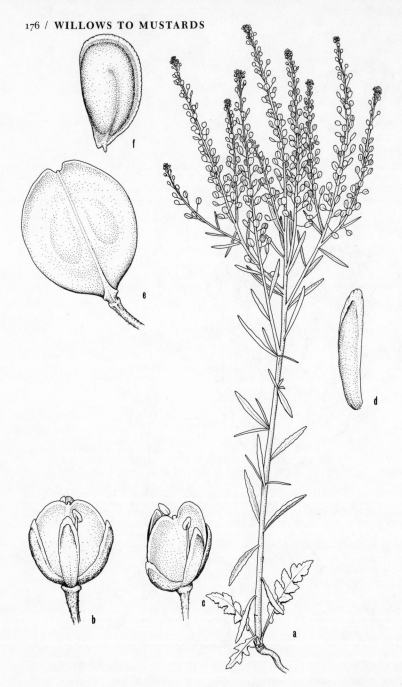

78. Lepidium densiflorum (Peppergrass). *a.* Habit, X½. *b, c.* Flowers, X17½. *d.* Sepal, X30. *e.* Fruit, X15. *f.* Seed, X25.

6. Lepidium densiflorum Schrad. Ind. Sem. Goett. 4. 1835.
Fig. 78.

Lepidium intermedium Gray, Pl. Wright. 2:15. 1853.

Lepidium neglectum Thell. Bull. Herb. Boiss. ser. 2, 4:708. 1904.

Annual or biennial herb from an elongated root; stems erect, branched, to 50 cm tall, glabrous; basal leaves pinnatifid, glabrous, to 10 cm long, long-petiolate, the ultimate segments entire to coarsely toothed; cauline leaves linear, entire to dentate, glabrous, to 2.5 cm long; racemes terminal, densely flowered, the flowers greenish, 1.0–1.5 mm broad, on spreading to ascending pedicels; sepals 4, linear, glabrous, green, 1.0–1.5 mm long; petals absent; stamens 2; silicle orbicular, emarginate, flat, unwinged, glabrous, 2.0–2.5 mm long, the minute, persistent style in the notch, the pedicels spreading, to 5 mm long.

COMMON NAME: Peppergrass.

HABITAT: Fields, pastures, roadsides, along railroads.

RANGE: Native of Europe and Asia; naturalized in most parts of North America.

ILLINOIS DISTRIBUTION: Occasional throughout the state, although less common in the southern counties.

This is the most common of the apetalous species of *Lepidium* in Illinois. It differs from *L. ruderale* by its once-pinnatifid basal leaves and its generally odorless leaves.

Illinois botanists during the nineteenth century called this species *L. intermedium* Gray.

It flowers from April to October.

15. Armoracia Gaertn., Mey. & Scherb. –Cress

Perennial herbs from rhizomes or deep roots; leaves finely dissected to merely toothed; inflorescence racemose, with showy flowers; flowers actinomorphic, white, bractless; sepals 4, free; petals 4, free; stamens 6; pistil one, the ovary superior, the style short or slender; silicles not flattened, tipped by the persistent style, the valves nerveless, the seeds few, arranged in two rows in each cell.

I am including *Neobeckia* Greene within *Armoracia*. The combined genera total about six species native to eastern North America, Europe, and Asia.

KEY TO THE SPECIES OF Armoracia IN ILLINOIS

1. Basal leaves divided into filiform divisions; aquatic plants _ _ _ _ _ _ _ _ _ _ _
_ _1. *A. aquatica*
1. Basal leaves not divided into filiform divisions; terrestrial plants _ _ _ _ _ _
_ _2. *A. lapathifolia*

1. **Armoracia aquatica** (Eat.) Wieg. Rhodora 27:186. 1925. *Fig.*
79.
Cochlearia aquatica Eat. Man. Bot. ed. 5, 181. 1829.
Nasturtium natans DC. var. *americanum* Gray, Ann. Lyc. N. Y.
3:223. 1836.
Nasturtium lacustre Gray, Gen. Fl. Am. Ill. 1:132. 1848.
Rorippa americana (Gray) Britt. Mem. Torrey Club 5:169. 1894.
Neobeckia aquatica (Eat.) Greene, Pittonia 3:95. 1896.
Radicula aquatica (Eat.) Robins. Rhodora 10:32. 1908.
Rorippa aquatica (Eat.) Palmer & Steyerm. Ann. Mo. Bot.
Gard. 22:550. 1935.

Aquatic perennial from slender rhizomes; stems much branched to simple, glabrous, to 75 cm long; basal immersed leaves to 15 cm long, long-petiolate, 1- to 5-pinnate, the ultimate segments capillary to filiform, entire, glabrous; emersed leaves lanceolate to oblong, obtuse to acute at the apex, cuneate to the sessile base, to 8 cm long, glabrous, entire to serrate; racemes lax, spreading or pendulous, few- to several-flowered, the flowers 5–10 mm broad, bractless; sepals linear to linear-lanceolate, green, glabrous, 2–4 mm long; petals 4, white, 5–9 mm long; silicles ovoid, not flat, glabrous, 1-locular, to 5 mm long, tipped by the persistent elongated style and the bilobed stigma, on spreading to ascending, glabrous pedicels to 1 cm long, with few small, plump, wingless seeds.

COMMON NAME: Lake Cress.
HABITAT: Swamps; quiet streams.
RANGE: Quebec to Ontario, south to Texas and Florida.
ILLINOIS DISTRIBUTION: Occasional throughout the state, but less common in the southern counties.
This strikingly unusual aquatic species, with its capillary-divided immersed basal leaves, is unlike any other member of the Brassicaceae in appearance.
A look at the synonymy will reveal that botanists have had difficulty deciding the systematic position of this species. It has been placed in no less than six different genera. Its relationship with *A. lapathifolia* is not at first obvious.
The lake cress flowers from May to August.

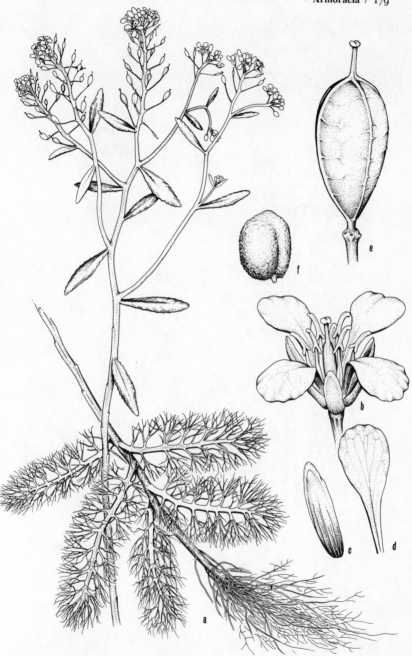

79. *Armoracia aquatica* (Lake Cress). *a*. Habit, X½. *b*. Flower, X5. *c*. Sepal, X7½. *d*. Petal, X5. *e*. Fruit, X7½. *f*. Seed, X15.

80. Armoracia lapathifolia (Horseradish). *a.* Upper part of plant, X½. *b.* Pinnatifid leaf, X1. *c.* Flower, X5. *d.* Sepal, X5. *e.* Petal, X5. *f.* Fruit, X12½.

2. Armoracia lapathifolia Gilib. Fl. Lith. 2:53. 1781. *Fig. 80.*
Cochlearia armoracia L. Sp. Pl. 648. 1753.
Cochlearia rusticana Lam. Fl. Fr. 2:471. 1778.
Armoracia rusticana (Lam.) Gaertn., Mey. & Scherb. Fl. Wett.
2:426. 1800.
Nasturtium armoracia (L.) Fries, Fl. Scand. 65. 1835.
Radicula armoracia (L.) Robins. Rhodora 10:32. 1908.
Armoracia armoracia (L.) Cockerell in Daniels, Fl. Boulder,
Colo. 130. 1911.

Terrestrial perennial from a deep, thick root; stems erect, sometimes branched, glabrous, to 1 m tall; basal leaves oblong, obtuse at
the apex, rounded or subcuneate at the base, crenate or serrate,
rarely pinnatifid, glabrous, to 30 cm long, petiolate; cauline leaves
lanceolate, acute at the apex, cuneate at the base, crenate, or the
lower ones pinnatifid, glabrous, sessile; racemes terminal and from
the uppermost axils, many-flowered, the flowers 5–10 mm broad,
bractless, on slender, ascending pedicels; sepals 4, elliptic, green,
glabrous, 3–5 mm long; petals 4, white, 5–10 mm long; silicles
globose or nearly so, glabrous, 2-locular, 4–7 mm in diameter,
tipped by the persistent style, on slender, glabrous pedicels longer
than the silicle.

COMMON NAME: Horseradish.
HABITAT: Fields, along roads.
RANGE: Native of Europe; occasionally cultivated and
sometimes escaped.
ILLINOIS DISTRIBUTION: Occasional throughout the state.
Horseradish of commerce comes from the thick roots of
this species.
The large basal leaves persist long into the autumn. The
flowers are attractive for a weedy species.
Armoracia lapathifolia flowers from May to July.

16. *Nasturtium R. Br.* –Water Cress

Perennial herb; leaves pinnately compound; racemes terminal;
flowers actinomorphic, bractless; sepals 4, free; petals 4, free; stamens 6; pistil one, the ovary superior, the style stout; siliques cylindrical, the valves nerveless, with several wingless seeds arranged
in two rows in each cell.
Only the following species comprises the genus.

1. **Nasturtium officinale** R. Br. in Ait. Hort. Kew. ed. 2, 4:109. 1812. *Fig. 81.*

Sisymbrium nasturtium-aquaticum L. Sp. Pl. 657. 1753.

Nasturtium siifolium Reichenb. Fl. Germ. Excurs. 683. 1830.

Radicula nasturtium-aquaticum (L.) Britten & Rendle, Brit. Seed Plants 3. 1907.

Perennial herb from fibrous roots; stems creeping or floating, rather fleshy, glabrous, rooting at many of the nodes; leaves pinnately compound, petiolate, the leaflets 3–11, oblong to oval to nearly orbicular, obtuse at the apex, rounded at the sessile or short-petiolulate base, entire to dentate-undulate, glabrous, sometimes fleshy; racemes terminal and from the uppermost axils, the flowers 4–6 mm broad, bractless; sepals 4, elliptic to narrowly oblong, green, glabrous, 1.5–2.5 mm long; petals 4, white, 3–6 mm long; siliques linear, cylindric, arcuate to straight, glabrous, to 1.8 cm long, to 2.5 mm broad, the beak absent or up to 1.5 mm long, the pedicels spreading to ascending, glabrous, with several seeds arranged in two rows in each cell.

COMMON NAME: Water Cress.

HABITAT: Cool springs and branches.

RANGE: Throughout North America, South America, Europe, and Asia.

ILLINOIS DISTRIBUTION: Occasional and scattered throughout the state.

Water cress has been widely cultivated in the past as a choice salad plant. Most manuals report that it is native to Europe and Asia, having escaped from cultivation in other parts of the world. I would agree with Steyermark (1964) in questioning the introduced nature of this species in Illinois and Missouri.

In addition to its value as a food plant for man, this species is an important source of nutrition for wildlife.

The flowers are produced from April to October.

17. *Thlaspi* L. – Penny Cress

Annual or perennial herbs; leaves basal and cauline, toothed, some or all the cauline ones sagittate or clasping; racemes terminal; flowers white, actinomorphic, bractless; sepals 4, free; petals 4, free; stamens 6; silicles flattened contrary to the partition, the valves keeled and winged, with 2–8 wingless seeds in each cell.

81. Nasturtium officinale (Water Cress). *a.* Habit, X1. *b.* Flower, X7½. *c.* Sepal, X12½. *d.* Petal, X7½. *e.* Fruiting branch, X½. *f.* Fruit, X6. *g.* Seed, X30.

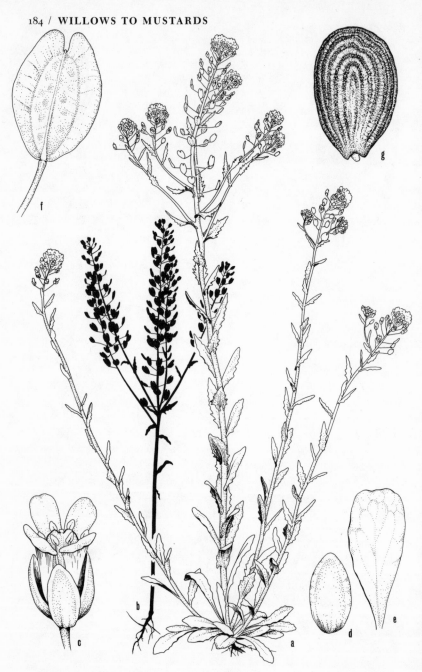

82. *Thlaspi arvense* (Field Penny Cress). *a*. Habit, X¼. *b*. Habit (in silhouette),
with leaves wilting and falling, X1/6. *c*. Flower, X10. *d*. Sepal, X15. *e*. Petal,
X10. *f*. Fruit, X2½. *g*. Seed, X20.

Thlaspi is composed of about two dozen species native to the temperate and arctic regions of the world.

KEY TO THE SPECIES OF Thlaspi IN ILLINOIS

1. Lowest stem leaves petiolate; seeds rugose, black, at least 2 mm long _____ 1. *T. arvense*
1. Lowest stem leaves sessile and clasping; seeds smooth, yellow-brown, 1.0–1.5 mm long _____ 2. *T. perfoliatum*

1. Thlaspi arvense L. Sp. Pl. 646. 1753. *Fig. 82.*

Annual herb from a thickened, elongated root; stems erect, branched or unbranched, glabrous, to 60 cm tall; basal leaves oblanceolate to narrowly obovate, acute at the apex, petiolate at the base, sparsely dentate, glabrous, withering during the spring; cauline leaves lanceolate to oblong, acute at the apex, sagittate at the clasping base, entire to dentate, glabrous; racemes terminal and from the upper axils, several-flowered, the flowers 2–3 mm broad, on spreading or ascending, glabrous pedicels; sepals 4, ovate, green, glabrous, 1–2 mm long; petals 4, white, 3–4 mm long, sometimes slightly emarginate; silicles orbicular, flat, deeply notched, broadly winged, to 15 mm long and broad, the style absent or minute, the seeds ovoid, somewhat flattened, black, rugose, 2.0–2.5 mm long.

COMMON NAME: Field Penny Cress.

HABITAT: Fields, roadsides, along railroads.

RANGE: Native of Europe and Asia; naturalized throughout North America.

ILLINOIS DISTRIBUTION: Scattered in all parts of the state. The seeds of the field penny cress have a peppery taste like those of *Lepidium*.

Thlaspi arvense is a fast-spreading weed in Illinois. Apparently the first reference to it from Illinois was made by Gates in 1926.

The flowers open from April to June.

2. Thlaspi perfoliatum L. Sp. Pl. 646. 1753. *Fig. 83.*

Annual herb from a thickened root; stems erect, usually unbranched, glabrous, to 30 cm tall; basal leaves ovate to ovate-lanceolate, acute to subacute at the apex, petiolate to sessile, sometimes cordate, sparsely dentate, glabrous, withering during the spring; cauline leaves oblong to lanceolate, acute at the apex, auriculate and clasping the stem at the base, glabrous, entire or nearly

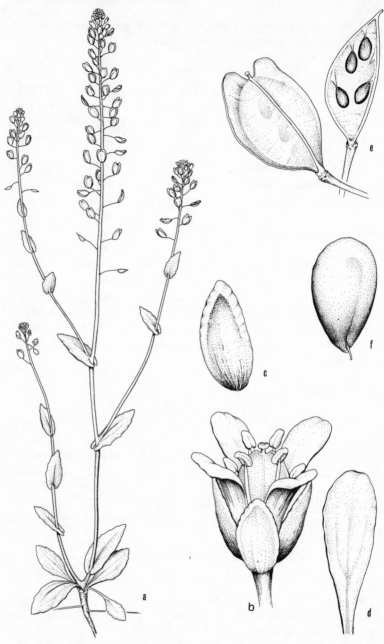

83. *Thlaspi perfoliatum* (Perfoliate Penny Cress). *a.* Habit, X½. *b.* Flower, X12½. *c.* Sepal, X20. *d.* Petal, X17½. *e.* Fruits, X7½. *f.* Seed, X20.

so, to 2.5 cm long, up to one-half as broad; racemes terminal and from the upper axils, several-flowered, the flowers 1.5–2.5 mm broad, on mostly spreading, glabrous pedicels up to 1.5 cm long; sepals 4, lanceolate, green, glabrous, 1–2 mm long; petals 4, white, 2–3 mm long; silicles obovoid to orbicular, flat, broadly notched, narrowly winged, to 6 mm long, often nearly as broad, the style minute, the seeds ovoid, somewhat flattened, yellow-brown, smooth, 1.0–1.5 mm long.

COMMON NAME: Perfoliate Penny Cress.

HABITAT: Waste ground.

RANGE: Native of Europe; adventive in several places in the eastern United States.

ILLINOIS DISTRIBUTION: Known only from Effingham and Shelby counties.

This uncommon adventive differs from the more abundant *T. arvense* by its smaller fruits and entire cauline leaves.

The perfoliate penny cress flowers during April and May.

18. *Cardaria Desv.* – Hoary Cress

Perennial herbs; leaves cauline, alternate, toothed, sessile or clasping; racemes borne in corymbs; flowers white, actinomorphic, bractless; sepals 4, free; petals 4, free; stamens 6; silicles inflated, more or less indehiscent, the valves 1-nerved or nerveless, with 2–4 wingless seeds per cell.

Cardaria is sometimes included within *Lepidium*, but differs primarily by its more or less indehiscent, unnotched silicles.

Only the following species occurs in Illinois.

1. **Cardaria draba** (L.) Desv. Journ. Bot. Desv. 3:163. 1814.
Fig. 84.

Lepidium draba L. Sp. Pl. 645. 1753.

Perennial herb from a rhizome; stems erect or ascending, branched, hoary-pubescent, to 45 cm tall; basal leaves oblong, obtuse to subacute at the apex, cuneate to the petiolate base, dentate, hoary-pubescent, to 10 cm long; cauline leaves similar but sagittate-clasping at the base; racemes in terminal corymbs, many-flowered, the flowers 2.5–4.0 mm broad, on ascending to spreading, filiform pedicels; sepals 4, lanceolate, green, glabrous or nearly so, 2–3 mm long; petals 4, white, 3–4 mm long; silicles broadly ovoid, inflated, glabrous, papillose, 3–4 mm long, unnotched, tipped by the persistent slender style, the valves papillose, keeled, with 2–4 wingless seeds.

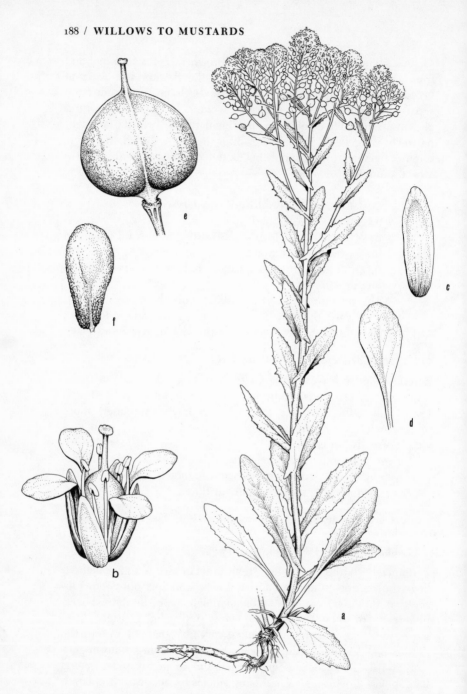

84. Cardaria draba (Hoary Cress). *a*. Habit, X⅓. *b*. Flower, X6. *c*. Sepal,
X12½. *d*. Petal, X10. *e*. Fruit, X5. *f*. Seed, X25.

COMMON NAME: Hoary Cress.

HABITAT: Fields, roadsides, along railroads.

RANGE: Native of Europe and Asia; naturalized in most parts of North America.

ILLINOIS DISTRIBUTION: Occasional in the northern half of the state, rare in the southern half.

This is a handsome weedy species, with the showy flowers arranged in corymbs at the tip of the stems. The flowers are mildly fragrant.

The flowers open from April to June.

19. *Alliaria* Scop. – Garlic Mustard

Onion-smelling biennial or perennial herbs; leaves alternate, toothed, cordate; racemes terminal; flowers white, actinomorphic, bractless; sepals 4, free, caducous; petals 4, free; stamens 6; pistil one, the ovary superior, the style short; siliques cylindric, terete, dishiscent, the valves 3-nerved, keeled, the seeds arranged in one row in each cell.

About five species native to Europe and Asia comprise the genus.

Only the following species occurs in Illinois.

1. **Alliaria officinalis** Andrz. ex DC. Syst. Veg. 2:489. 1821. *Fig. 85.*

Erysimum alliaria L. Sp. Pl. 660. 1753.

Sisymbrium alliaria (L.) Scop. Fl. Carn., ed. 2, 2:26. 1772.

Alliaria alliaria (L.) Britt. Mem. Torrey Club 5:167. 1894.

Perennial herb; stems erect, branched, to 1 m tall, glabrous or sparsely pubescent; lower leaves reniform to broadly ovate, sub-acute to acute at the apex, truncate to cordate at the base, coarsely dentate or crenate or undulate, to 1.5 cm long, glabrous or sparsely pubescent, on long petioles; upper leaves similar but on short pet-ioles or nearly sessile; racemes terminal, several-flowered, the flow-ers 6–10 mm broad, bractless; sepals 4, oval, green, glabrous, 1–3 mm long, caducous; petals 4, white, oblong, clawed, 5–9 mm long; siliques linear, cylindric, terete, becoming 4-angled when dry, gla-brous, to 5 cm long, the persistent style minute, the pedicels short, stout, spreading, the seeds oblongoid, arranged in one row in each locule.

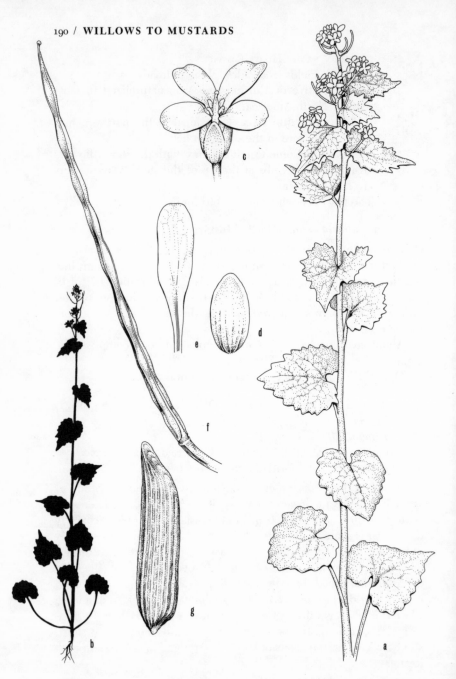

85. *Alliaria officinalis* (Garlic Mustard). *a.* Upper part of plant, X¼. *b.* Habit (in silhouette), X⅛. *c.* Flower, X5. *d.* Sepal, X10. *e.* Petal, X6. *f.* Fruit, X2½. *g.* Seed, X25.

COMMON NAME: Garlic Mustard.

HABITAT: Disturbed woods.

RANGE: Native of Europe and Asia; adventive in the eastern half of the United States; also Quebec and Ontario.

ILLINOIS DISTRIBUTION: Confined to the northern half of Illinois.

The alliaceous odor imparted by this species is unique in this family which is composed generally of scentless species.

Although this species is not particularly common in Illinois, it is abundant at Allerton Park, Piatt County.

Patman and Iltis (1961) have given reasons for calling this species *A. petiolata*.

The flowers bloom in May and June.

20. *Berteroa DC.* – Hoary Alyssum

Annual or perennial herbs with stellate pubescence; leaves cauline, alternate, entire or toothed; racemes terminal; flowers white, actinomorphic, bractless; sepals 4, free; petals 4, free, 2-cleft; stamens 6; silicles flat to somewhat inflated, the slender style persistent, the seeds several in each cell, winged.

Berteroa is composed of five species native to Europe and Asia. It differs from *Alyssum* by its bifid petals.

Only the following species occurs in Illinois.

1. **Berteroa incana** (L.) DC. Syst. 2:291. 1821. *Fig. 86.*
Alyssum incanum L. Sp. Pl. 650. 1753.

Perennial herbs from elongated roots; stems ascending to erect, branched, hoary-pubescent, gray-green, to 1 m tall; leaves lanceolate to oblong, subacute to obtuse at the apex, cuneate to the base, entire or undulate, hoary-pubescent, the lowermost short-petiolate; racemes terminal, many-flowered, the flowers 2.0–3.5 mm broad, on ascending pedicels; sepals 4, broadly lanceolate, 1.0–1.5 mm long; petals 4, white, bifid, 2.5–3.5 mm long; silicles swollen, oblongoid, 5–9 mm long, 2.5–3.5 mm thick, canescent-pubescent, the persistent style 2–3 mm long, the pedicels ascending, 4–7 mm long, the seeds several, winged.

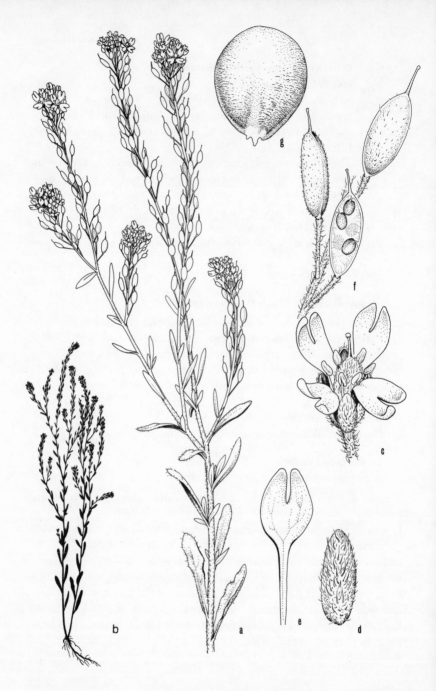

86. *Berteroa incana* (Hoary Alyssum). *a*. Upper part of plant, X½. *b*. Habit (in silhouette), X⅛. *c*. Flower, X7½. *d*. Sepal, X20. *e*. Petal, X12½. *f*. Fruits, X2½. *g*. Seed, X25.

COMMON NAME: Hoary Alyssum.

HABITAT: Fields, roadsides, along railroads.

RANGE: Native of Europe; naturalized in the northeastern United States.

ILLINOIS DISTRIBUTION: Not common in the northern half of the state; also Cumberland and St. Clair counties.

The hoary pubescence of this species gives it a distinctive gray-green appearance. The bifid petals and the plump, canescent-pubescent silicles are additional field characteristics.

Swink (1974) has observed a considerable increase of this species in northeastern Illinois in recent years.

The flowers are produced from May to September.

21. *Lobularia Desv.* – Sweet Alyssum

Perennial herbs (in Illinois) or shrubs, usually with forked hairs; leaves alternate, entire; racemes terminal; flowers white, actinomorphic, bractless; sepals 4, free; petals 4, free; stamens 6, silicles flattened, the valves nerveless, the slender style persistent, the seeds one in each cell.

There are four species of *Lobularia*, all native to the Mediterranean region.

Only the following species occurs in Illinois.

1. **Lobularia maritima** (L.) Desv. Journ. Bot. Desv. 3:169. 1814. *Fig. 87*.

Clypeola maritima L. Sp. Pl. 652. 1753.

Alyssum maritimum Lam. Encycl. 1:98. 1783.

Koniga maritima R. Br. in Denh. & Clapp, Narr. Exp. Afric. 214. 1826.

Perennial herb from an elongated, thickened root; stems spreading to ascending, much branched, appressed-pubescent, to 30 cm long; basal leaves oblanceolate, obtuse to subacute at the apex, tapering to the short-petiolate base, entire, minutely pubescent; cauline leaves linear to narrowly lanceolate, acute at the apex, cuneate to the nearly sessile base, minutely pubescent, to 8 cm long, to 5 mm wide; racemes terminal and from the upper leaf axils, many-flowered, the flowers fragrant, 3.5–5.0 mm broad, on ascending pedicels; sepals 4, elliptic, glabrous or pubescent, caducous, 1–2 mm long; petals 4, white, orbicular to obovate, long-clawed, 3–4 mm long; silicles oval to orbicular, somewhat flattened, glabrous, tipped by the persistent style, 2.5–3.5 mm long, the pedicels spreading to

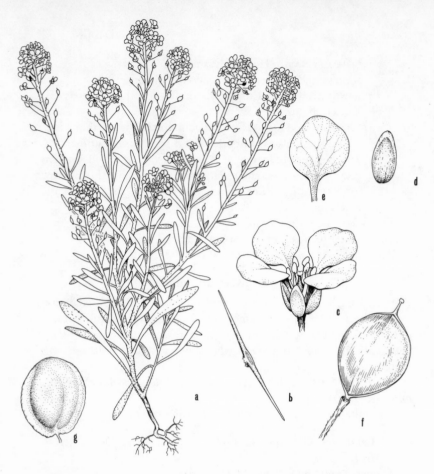

87. *Lobularia maritima* (Sweet Alyssum).   *a*. Habit, X½.   *b*. Two-parted hair, X100.   *c*. Flower, X5.   *d*. Sepal, X10.   *e*. Petal, X6.   *f*. Fruit, X7½.   *g*. Seed, X25.

ascending, to nearly 10 mm long, the seeds wingless or slightly winged.

COMMON NAME: Sweet Alyssum.

HABITAT: Waste ground.

RANGE: Native of Europe; cultivated as a garden ornamental and sometimes escaped in several states of the northern United States.

ILLINOIS DISTRIBUTION: Known from Champaign, Cook, Du Page, Hancock, Lake, and McHenry counties.

Lobularia maritima is the common garden species known as sweet alyssum. It is similar to the genus *Alyssum*, but has bifurcate rather than stellate hairs and white rather than yellow flowers.

The flowers bloom from June to August.

22. *Barbarea R. Br.* – Winter Cress

Biennial or perennial herbs from elongated rootstocks; leaves pinnate to pinnatifid to lobed, the basal in overwintering rosettes, the cauline sometimes with clasping bases; racemes terminal and from the upper axils; flowers yellow, actinomorphic, bractless; sepals 4, free; petals 4, free; stamens 6; siliques terete or 4-angled, the valves keeled along the midnerve, the short style persistent, the seeds flat, wingless, arranged in one row in each cell.

Barbarea has about six species native to most temperate regions of the world. The genus is not easily segregated from *Brassica* on obvious characters.

Barbarea species are usually glabrous, while *Brassica* species are usually pubescent, except for *Brassica juncea*, *B. rapa*, *B. oleracea*, and *B. napus*. *Barbarea* species have auriculate-clasping cauline leaves, while *Brassica* species usually do not, except for *B. rapa*, *B. oleracea*, and *B. napus*. If characters of the pubescence and leaf bases fail, then the beak of the silique must be examined, for *Barbarea* has small beaks less than 5 mm long, while *Brassica* has long beaks more than 5 mm long.

KEY TO THE SPECIES OF Barbarea IN ILLINOIS

1. All leaves pinnatifid; beak of silique up to 1 mm long _____ 1. *B. verna*
1. Uppermost leaves merely dentate; beak of silique 1.5–3.0 mm long _____2. *B. vulgaris*

1. **Barbarea verna** (Mill.) Aschers. Fl. Prov. Brandenb. 1:36. 1864. *Fig. 88.*

Erysimum vernum Mill. Gard. Dict. ed. 8, no. 3. 1768.

Tufted biennial from elongated, thickened roots; stems erect, branched, glabrous, to 75 cm tall; all leaves similar, pinnatifid with 10–20 lobes, the terminal lobe the largest, oval to obovate, entire or repand, glabrous, the lowest leaves petiolate, the upper auriculate-clasping; racemes terminal and from the axils of the uppermost leaves, several-flowered, the flowers 6–8 mm broad, bractless, on ascending pedicels; sepals 4, narrowly lanceolate, green, glabrous, 2.5–4.0 mm long; petals 4, bright yellow, spatulate, 6–8 mm long; siliques 4-angled, somewhat compressed, rigid, glabrous, to 8 cm long including the 0.5–1.0 mm long beak, the pedicels stout, as-

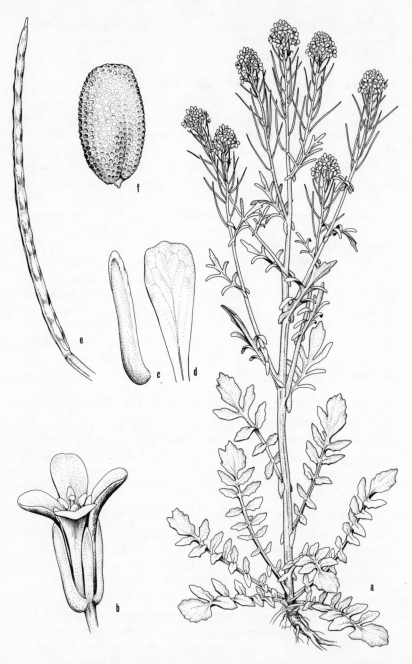

88. *Barbarea verna* (Early Winter Cress). *a.* Habit, X¼. *b.* Flower, X7½. *c.* Sepal, X10. *d.* Petal, X6. *e.* Fruit, X1. *f.* Seed, X20.

cending, to 8 mm long, the seeds oblongoid, gray-brown, reticulate, to 1.5 mm long.

COMMON NAME: Early Winter Cress.

HABITAT: Waste ground.

RANGE: Native of Europe; sparsely adventive throughout the United States.

ILLINOIS DISTRIBUTION: Known first from Johnson County (field, May 9, 1954, *H. E. Ahles 7997*), more recently from Champaign and Massac counties.

The winter cress derives its common name from the fact that the basal rosette of leaves overwinters.

This species is similar to the common *B. vulgaris*, but differs in having all its leaves pinnatifid.

The flowers open from April to June.

2. Barbarea vulgaris R. Br. in Ait. Hort. Kew. ed. 2, 4:109. 1812.

Erysimum barbarea L. Sp. Pl. 660. 1753.

Tufted biennial from elongated, thick roots; stems erect, branched, glabrous, to 75 cm tall; basal leaves pinnatifid with 3–7 (–9) lobes, petiolate, the terminal lobe the largest, oval to obovate, entire to repand, glabrous, the leaves to 12 cm long; lowermost cauline leaves similar to the basal leaves, uppermost cauline leaves obovate to orbicular, shallowly lobed, dentate, or entire, obtuse at the apex, auriculate-clasping at the base, glabrous; racemes terminal and from the axils of the uppermost leaves, several-flowered, the flowers 6–8 mm broad, bractless, on ascending to erect pedicels; sepals 4, narrowly lanceolate, green, glabrous, 2.5–4.0 mm long; petals 4, bright yellow, spatulate, 6–8 mm long; siliques scarcely 4-angled, cylindrical, rigid, glabrous, to 3 cm long including the 1.5–3.0 mm long beak, the pedicels slender, erect to ascending to spreading, to 6 mm long, the seeds oblongoid, gray, rugulose, to 1.5 mm long.

Two varieties occur in Illinois.

KEY TO THE VARIETIES OF Barbarea vulgaris IN ILLINOIS

1. Siliques erect, closely appressed to the stem _____
_____2a. *B. vulgaris* var. *vulgaris*
1. Siliques spreading to ascending _____ 2b. *B. vulgaris* var. *arcuata*

2a. Barbarea vulgaris (L.) R. Br. var. **vulgaris** *Fig. 89a–g.*

Barbarea barbarea (L.) MacM. Met. Minn. 259. 1892.

Siliques erect, closely appressed to the stem.

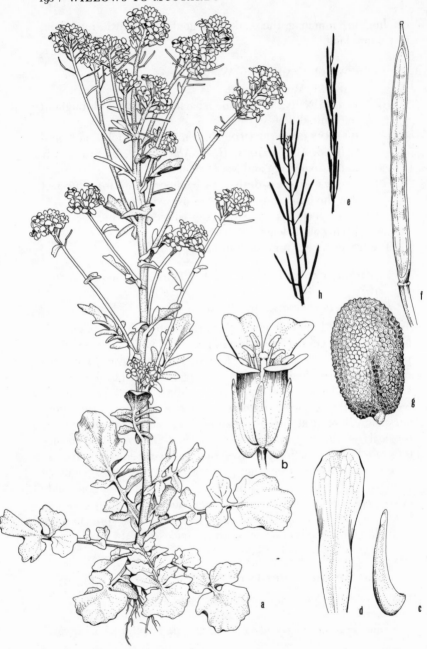

89. *Barbarea vulgaris* (Yellow Rocket). *a.* Habit, X¼. *b.* Flower, X5. *c.* Sepal, X10. *d.* Petal, X6. *e.* Fruiting raceme (in silhouette), X⅓. *f.* Fruit, X2½. *g.* Seed, X20. var. *arcuata.* *h.* Fruiting raceme (in silhouette), X⅓.

COMMON NAMES: Yellow Rocket; Winter Cress.

HABITAT: Fields.

RANGE: Native of Europe; naturalized in the eastern United States.

ILLINOIS DISTRIBUTION: Known only from Jackson County.

The typical variety is apparently very rare in Illinois. It is sometimes confused with *B. stricta* Andrz., which is a different species. It is distinguished from the very common var. *arcuata* by the erect, appressed siliques.

The flowers open from April to June.

2b. Barbarea vulgaris (L.) R. Br. var. **arcuata** (Opiz) Fries, Novit. Fl. Suec. 205. 1828. *Fig. 89h.*

Erysimum arcuatum Opiz ex J. & C. Presl, Fl. Cechica 138. 1819.

Siliques spreading to ascending.

COMMON NAMES: Yellow Rocket; Winter Cress.

HABITAT: Fields, pastures, along roads.

RANGE: Native of Europe; naturalized in eastern North America.

ILLINOIS DISTRIBUTION: Common; in every county.

This variety of *B. vulgaris* is common in all parts of the state, where it occurs in a wide variety of disturbed habitats.

The leaves are very shiny on the upper surface. The flowers are extremely showy.

The young leaves and stems can be cooked for vegetable greens. I am not convinced that this variety with spreading siliques is varietally distinct from var. *vulgaris*.

The flowers bloom from April to June.

23. *Lesquerella* S. Wats. – Bladder Pod

Annual or perennial stellate-pubescent herbs from fibrous roots, taproots, or thickened caudices; leaves simple, basal and alternate; racemes terminal; flowers yellow, actinomorphic, bractless; sepals 4, free; petals 4, free; stamens 6; silicles inflated, the valves nerveless, the partition hyaline and nerved, the seeds flat, narrowly winged or unwinged, arranged in two rows in each cell.

Lesquerella is primarily a North American genus of about 35 species. Most of them occur in the western United States. Rollins and Shaw (1973) have monographed the genus.

KEY TO THE SPECIES OF Lesquerella IN ILLINOIS

1. Plants perennial; fruits densely stellate-pubescent _____
_____ 1. *L. ludoviciana*
1. Plants annual; fruits glabrous or nearly so _____ 2. *L. gracilis*

1. Lesquerella ludoviciana (Nutt.) S. Wats. Proc. Am. Acad.
23:252. 1888. *Fig. 90.*
Alyssum ludovicianum Nutt. Gen. Pl. 2:63. 1818.

Tufted perennial herb from a rather stout caudex; stems decumbent
to ascending, simple to sparingly branched, densely stellate-pubescent, to 40 cm tall; leaves linear to oblanceolate, obtuse to subacute
at the apex, cuneate to the base, entire to repand, densely stellate-
pubescent, scabrous, the basal to 10 cm long, the cauline smaller;
racemes terminal and from the axils of the uppermost leaves, several-flowered, the flowers to 15 mm broad, bractless, on spreading
or recurved pedicels; sepals 4, lanceolate, stellate-pubescent, 4–7
mm long; petals 4, yellow, spatulate, 6–10 mm long; silicles globose
to oval, inflated, densely stellate-pubescent, stipitate, 3–4 mm in
diameter, the persistent style slender, 3–4 mm long, on slender,
recurved pedicels 1–2 cm long.

COMMON NAME: Silvery Bladder Pod.
HABITAT: Sandy soil.
RANGE: Minnesota to Manitoba and Montana, south to
Arizona and Kansas; northwestern Illinois.
ILLINOIS DISTRIBUTION: Known only from Mason County
(original collection 12 miles northeast of Havana, August
22, 1904, *H. A. Gleason*). This species still exists in this
area.
This is one of the several rare and disjunct species found
in the sand prairies near Havana in Mason County. The
nearest known station is many miles away.

This species was reported by Gleason (1910) and others as *L.
argentea* (Pursh) Mac M., but this is not the same species that Sereno
Watson had named earlier as *L. argentea*.

The silvery bladder pod flowers from May to August.

2. Lesquerella gracilis (Hook.) S. Wats. Proc. Am. Acad. 23:
253. 1888. *Fig. 91.*
Vesicaria gracilis Hook. Bot. Mag. 63:Pl. 3533. 1836.

Annual from fibrous roots; stems decumbent to erect, simple or
branched, slender, sparsely stellate-pubescent, to 50 cm long; leaves

90. *Lesquerella ludoviciana* (Silvery Bladder Pod). *a.* Habit, X½. *b.* Flower, X5.
c. Sepal, X7½. *d.* Petal, X6. *e.* Pistil and stamen, at anthesis, X7½. *f.* Fruit,
X5. *g.* Seed, X25.

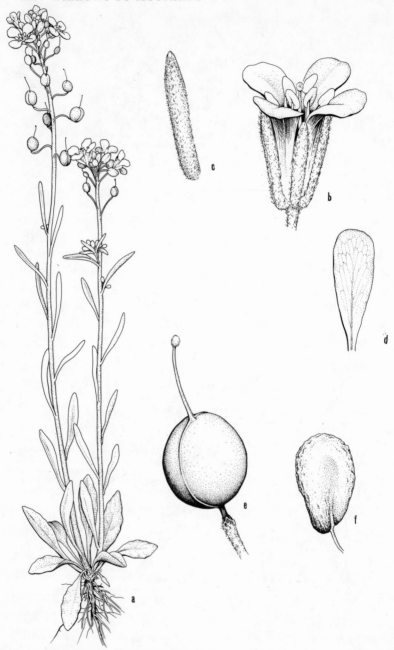

91. *Lesquerella gracilis* (Slender Bladder Pod). *a.* Habit, X½. *b.* Flower, X6.
c. Sepal, X10. *d.* Petal, X6. *e.* Fruit, X7½. *f.* Seed, X20.

linear to oblanceolate, subacute at the apex, cuneate to the base, entire or repand, sparsely stellate-pubescent, the basal to 8 cm long and short-petiolate, the cauline to 4.5 cm long and sessile; racemes terminal and from the axils of the upper leaves, several-flowered, the flowers 5–8 mm broad, bractless, on recurved to spreading pedicels; sepals 4, lanceolate, green, glabrous or pubescent, 3–5 mm long; petals 4, yellow, narrowly obovate, 5–7 mm long; silicles globose, inflated, glabrous or nearly so, stipitate, 3–4 mm in diameter, the persistent style slender, 3–4 mm long, on slender recurved pedicels, 1–2 cm long.

COMMON NAME: Slender Bladder Pod.
HABITAT: Along railroad (in Illinois).
RANGE: Iowa south to Oklahoma and Texas; adventive eastward.
ILLINOIS DISTRIBUTION: Known only from Cook County.
Only two collections from Cook County are known for this adventive species in Illinois, both made on June 9, 1894.

24. Diplotaxis DC. – Rocket

Annual, biennial, or perennial herbs; leaves basal and cauline, pinnatifid or toothed; racemes terminal; flowers yellow or white, actinomorphic, bractless; sepals 4, free; petals 4, free; stamens 6; siliques linear, flat, the valves 1-nerved, the seeds wingless, arranged in two rows in each cell.

About twenty species of *Diplotaxis* are native to Europe and Asia. The genus is similar to *Brassica*, differing mainly by its seeds arranged in two rows in each cell and by its flattened siliques.

KEY TO THE SPECIES OF Diplotaxis IN ILLINOIS

1. Sepals 5–8 mm long; stems leafy throughout; fruits stipitate _____ _____1. *D. tenuifolia*
1. Sepals 2.5–5.0 mm long; stems leafy only in the lower half; fruits sessile _____ 2. *D. muralis*

1. Diplotaxis tenuifolia (L.) DC. Syst. 2:632. 1753. *Fig. 92.*
Sisymbrium tenuifolium L. Cent. Pl. 1:18. 1755.

Perennial herb from thickened roots; stems erect, much branched, leafy throughout, more or less glabrous, glaucous, to 1 m tall; leaves deeply pinnatifid, acute at the apex, petiolate at the base, glabrous

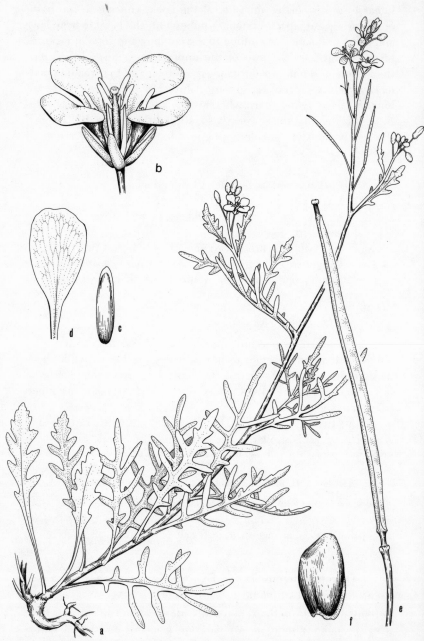

92. *Diplotaxis tenuifolia* (Sand Rocket). *a.* Habit, X⅓. *b.* Flower, X2½. *c.* Sepal, X3½. *d.* Petal, X2½. *e.* Fruit, X2½. *f.* Seed, X20.

or nearly so, the largest to 15 cm long; racemes terminal, several-flowered, the flowers 1.5–2.0 cm broad, bractless, on erect pedicels; sepals 4, broadly lanceolate, green, glabrous, 5–8 mm long; petals 4, yellow, clawed, 10–20 mm long; siliques linear, glabrous, to 5 cm long, stipitate, the slender, erect, glabrous pedicels to 3 cm long, the seeds wingless.

COMMON NAME: Sand Rocket.

HABITAT: Along railroad (in Illinois).

RANGE: Native of Europe; naturalized in eastern North America.

ILLINOIS DISTRIBUTION: Cook County (Brainard Avenue, July 2, 1957, *Glassman 4396, Chapp 124*).

Diplotaxis tenuifolia is distinguished from *D. muralis* by its longer sepals and stipitate fruits.

The flowers open from July to September.

2. Diplotaxis muralis (L.) DC. Syst. 2:634. 1821. *Fig. 93.*
Sisymbrium murale L. Sp. Pl. 658. 1753.

Annual herb from fibrous roots; stems erect, branched, leafy only in the lower half, glabrous or sparsely hispid, to 75 cm tall; leaves pinnatifid or coarsely toothed, subacute to acute at the apex, petiolate at the base, usually glabrous, the largest to 10 cm long; racemes terminal, several-flowered, the flowers 1.2–1.7 cm broad, bractless, on spreading to ascending pedicels; sepals 4, lanceolate, green, glabrous, 3–5 mm long; petals 4, yellow, clawed, 6–12 mm long; siliques linear, glabrous, to 3.5 cm long, not stipitate, erect but on spreading to ascending, glabrous pedicels to 1.5 cm long, the seeds wingless.

COMMON NAME: Wall Rocket.

HABITAT: Along railroads; in lawns.

RANGE: Native of Europe; naturalized in eastern North America.

ILLINOIS DISTRIBUTION: Not common; known only from the northern one-fourth of the state.

This species has less deeply divided leaves than does *D. tenuifolia*.

It flowers from June to August.

93. *Diplotaxis muralis* (Wall Rocket).   *a.* Habit, X⅓.   *b.* Flower, X10.   *c.* Sepal, X4.   *d.* Petal, X5.   *e.* Fruit, X2½.   *f.* Seed, X20.

25. *Erysimum* L.–Treacle Mustard

Annual, biennial, or perennial herbs, with bifurcate hairs; leaves basal and cauline, entire, toothed, or lobed; racemes terminal and from the axils of the uppermost leaves; flowers yellow or orange, actinomorphic, bractless; sepals 4, free; petals 4, free; stamens 6; stigmas lobed; siliques linear, 4-angled, the valves keeled along the midvein, the seeds wingless, arranged in one row in each cell.

Erysimum is a genus of about one hundred species native to the northern hemisphere. All Illinois species at one time were included in the segregate genus *Cheirinia* Link.

KEY TO THE SPECIES OF Erysimum IN ILLINOIS

1. Petals well over 1 cm long, orange-yellow _____ 1. *E. capitatum*
1. Petals up to 1 cm long, yellow _____ 2
  2. Petals up to 5 mm long _____ 2. *E. cheiranthoides*
  2. Petals 5–10 mm long _____ 3
3. Annual; fruit more than 5 cm long; plants pale green _____
    _____3. *E. repandum*
3. Perennial; fruit up to 5 cm long; plants gray-green _____
    _____4. *E. inconspicuum*

1. **Erysimum capitatum** (Dougl.) Green, Fl. Francisc. 269. 1891. *Fig. 94.*
Cheiranthus capitatus Dougl. in Hook. Fl. Bor. Am. 1:38. 1833.
Erysimum arkansanum Nutt. ex Torr. & Gray, Fl. N. Am. 1:95. 1838.
Erysimum asperum (Nutt.) DC. var. *arkansanum* (Nutt.) Patterson, Cat. Pl. Ill. 4. 1876.

Biennial or perennial herb from a stout root; stems erect, simple or branched, pubescent, to 1 m tall; leaves lanceolate, acute at the apex, cuneate at the base, entire or denticulate, pubescent, the basal leaves to 2 cm long, petiolate, the cauline leaves larger, sessile; racemes terminal, several-flowered, the flowers 1.2–2.5 cm broad, bractless, on stout, spreading pedicels; sepals 4, lanceolate, green, glabrous, 1.0–1.5 cm long; petals 4, orange-yellow, 1.2–2.5 cm long; siliques linear, 4-angled, more or less pubescent, to 10 cm long, thickly bilobed at the tip, the spreading pedicels to 8 mm long.

94. Erysimum capitatum (Western Wallflower). *a.* Habit, X¼. *b.* Flower, X2.
c. Sepal, X3. *d.* Petal, X2½. *e.* Fruiting raceme, X½. *f.* Fruit, X2. *g.* Seed,
X20.

COMMON NAME: Western Wallflower.

HABITAT: Sandy bluffs.

RANGE: Ohio to Washington, south to California, Texas, and Missouri.

ILLINOIS DISTRIBUTION: Confined to a few west-central counties; also Kendall and La Salle counties.

This is one of the showiest as well as rarest wildflowers in Illinois. The orange-yellow flowers bloom from May to July. This is also the only species of *Erysimum* in the state with petals in excess of one centimeter long.

I am following Rossbach (1958) in using the binomial *E. capitatum* for this species. Fernald (1950) employs the binomial *E. arkansanum*, while Gleason (1952) and Jones (1963) use *Erysimum asperum*. This last name applies to a different species, however.

Mead reported the existence of this species in Illinois as early as 1846.

2. **Erysimum cheiranthoides** L. Sp. Pl. 661. 1753. *Fig. 95.*
Cheirinia cheiranthoides (L.) Link, Enum. Hort. Berol. 2:170. 1820.

Annual herb from fibrous roots; stems erect or ascending, branched, minutely rough-pubescent, to 1 m tall; leaves lanceolate, subacute to acute at the apex, cuneate at the base, entire or nearly so, minutely pubescent, to 10 cm long, the lower leaves petiolate, the upper sessile; racemes terminal, several-flowered, the flowers 4–6 mm broad, bractless, on spreading to ascending pedicels; sepals 4, lanceolate, green, pubescent, 1–2 mm long; petals 4, yellow, 3–5 mm long; siliques linear, obscurely 4-angled, glabrous or minutely pubescent, to 3 cm long, minutely beaked by the persistent style, erect but on spreading-ascending pedicels, the seeds oblongoid, unwinged, brown, 0.8–1.1 mm long.

COMMON NAME: Wormseed Mustard.

HABITAT: Barnyards, fields, roadsides, along railroads.

RANGE: Native of Europe and apparently naturalized throughout North America.

ILLINOIS DISTRIBUTION: Occasional in the northern half of the state, absent in the southern half, except for Clark County.

There is some controversy as to whether this species is native or naturalized in North America. In Illinois, it is a weedy species of waste ground.

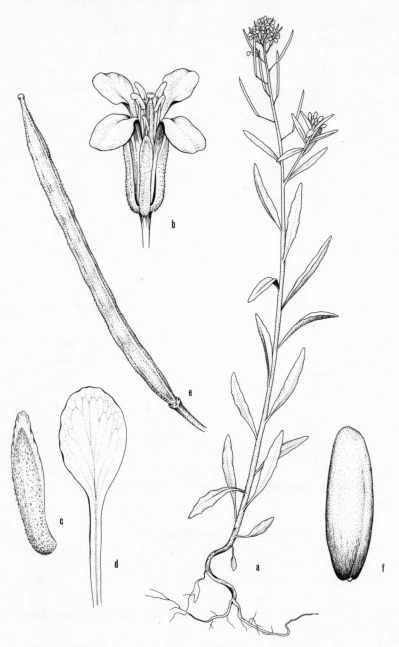

95. *Erysimum cheiranthoides* (Wormseed Mustard). *a*. Habit, X¼. *b*. Flower, X15. *c*. Sepal, X25. *d*. Petal, X15. *e*. Fruit, X4. *f*. Seed, X25.

Erysimum cheiranthoides has the smallest flowers of any *Erysimum* in Illinois. The flowers open from May to September.

3. **Erysimum repandum** L. Amoen. Acad. 3:415. 1756. *Fig. 96.*
Cheirinia repanda Link, Enum. Hort. Berol. 2:171. 1820.

Annual herb from fibrous roots; stems erect, simple to more often branched, pubescent, to 50 cm tall; leaves lanceolate, acute at the apex, cuneate at the base, repand-denticulate to sinuately lobed, pubescent on both surfaces, to 8 cm long, the basal often petiolate, the upper mostly sessile; racemes terminal, several-flowered, the flowers 7–10 mm broad, bractless, the pedicels ascending; sepals 4, lanceolate, green, 2–4 mm long, slightly unequal, the two longer ones strongly keeled; petals 4, pale yellow, 5–10 mm long; siliques linear, 4-angled, torulose, glabrous or strigose, 5–10 cm long, the persistent style minute, the pedicels stout, ascending, to 7 mm long, the seeds oblong, brown, about 1 mm long.

COMMON NAME: Treacle Mustard.
HABITAT: Along roads and railroads.
RANGE: Native of Europe; naturalized throughout the United States.
ILLINOIS DISTRIBUTION: Occasional and scattered throughout the state.
This species often grows densely along roads where it imparts a solid pale yellow color to the shoulders.
The flowers are intermediate in size between those of *E. capitatum* and *E. cheiranthoides*. They are similar to those of *E. inconspicuum*. *Erysimum repandum* differs from *E. inconspicuum* by its annual habit and its longer siliques.
The flowers appear from April to June.

4. **Erysimum inconspicuum** (S. Wats.) MacM. Met. Minn. Val. 268. 1892. *Fig. 97.*
Erysimum parviflorum Nutt. ex Torr. & Gray, Fl. N. Am. 1:95. 1838, non Pers. (1807).
Erysimum asperum DC. var. *inconspicuum* S. Wats. Bot. King's Exp. 24. 1871.

Perennial from an elongated root; stems erect, simple or branched, gray-pubescent, scabrous, to 50 cm tall; leaves linear to oblanceolate, obtuse to subacute at the apex, cuneate at the petiolate or sessile base, entire or sparsely repand-dentate, gray-pubescent on both surfaces, to 7.5 cm long; racemes terminal, several-flowered,

96. Erysimum repandum (Treacle Mustard). *a.* Habit, X⅓. *b.* Flower, X10. *c.* Sepal, X11. *d.* Petal, X11. *e.* Fruit, X1½. *f.* Seed, X32½.

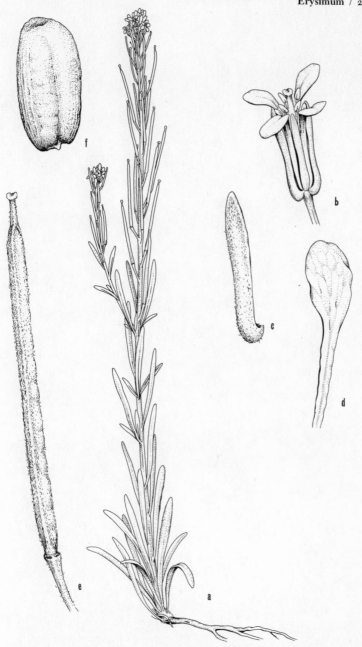

97. *Erysimum inconspicuum* (Small Wormseed Mustard). *a*. Habit, X⅓. *b*. Flower, X2½. *c*. Sepal, X5. *d*. Petal, X5. *e*. Fruit, X2½. *f*. Seed, X30.

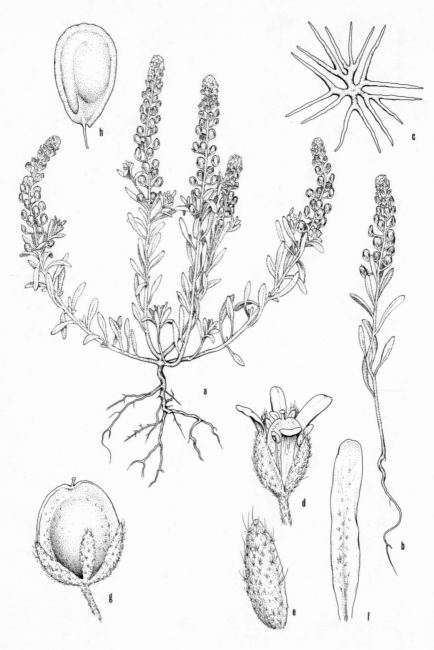

98. *Alyssum alyssoides* (Pale Alyssum). *a, b*. Habit, X½. *c*. Stellate hair, X100.
d. Flower, X15. *e*. Sepal, X30. *f*. Petal, X22½. *g*. Fruit, X10. *h*. Seed, X20.

the flowers 8–10 mm broad, bractless, on ascending to nearly erect pedicels; sepals 4, narrowly oblong, 6–8 mm long; petals 4, pale yellow, 7–10 mm long; siliques linear, 4-angled, gray-pubescent, scabrous, 2–5 cm long, the persistent style short and stout, the pedicels ascending to erect, to 5 mm long, the seeds oblong, brown, 1.0–1.3 mm long.

COMMON NAME: Small Wormseed Mustard.

HABITAT: Along railroads.

RANGE: Ontario to British Columbia, south to Nevada and South Dakota; adventive eastward in the United States.

ILLINOIS DISTRIBUTION: Occasionally found in the northern half of the state; also St. Clair County.

This species has a distinctive gray-green appearance. It is primarily a weed along railroads, where it has been introduced from the northwestern United States.

The flowers open in May and June.

26. Alyssum L.–Alyssum

Annual or perennial herbs with stellate pubescence; leaves mostly unlobed, alternate; racemes terminal; flowers yellow, actinomorphic, bractless; sepals 4, free; petals 4, free; stamens 6; silicle compressed, the valves nerveless, the seeds wingless, arranged in 1–2 rows in each cell.

Alyssum is a genus of about one hundred small herbs native to the Old World. It differs from Lobularia by its yellow flowers and stellate pubescence.

Only the following species occurs in Illinois.

1. **Alyssum alyssoides** (L.) L. Syst., ed. 10, 1130. 1759. *Fig. 98.*

Clypeola alyssoides L. Sp. Pl. 652. 1753.

Annual herb from fibrous roots; stems erect, branched from the base, hoary-pubescent, to 25 cm tall; leaves linear to spatulate, obtuse at the apex, cuneate at the base, entire, to 2.5 cm long, hoary-pubescent, the lowest leaves petiolate, the others sessile; racemes terminal, densely many-flowered, the flowers 1.5–2.5 mm broad, on spreading or ascending pedicels; sepals 4, ovate, about 1 mm long, persistent in fruit; petals 4, pale yellow, 1.5–2.5 mm long; silicles orbicular, sharply margined, minutely pubescent, to 3 mm in diameter, the persistent style minute, the pedicels spreading to ascending, to 5 mm long, the seeds 2 in each cell.

COMMON NAME: Pale Alyssum.
HABITAT: Along railroads.
RANGE: Native of Europe; naturalized in much of the United States.
ILLINOIS DISTRIBUTION: Occasional in the northern half of the state; also Crawford and St. Clair counties.
This dwarf, bushy, tufted annual flowers during May and June.
All parts of the plant are stellate-pubescent.

27. *Isatis* L. – Woad

Annual or biennial herbs; leaves simple, undivided; racemes arranged in a terminal corymb; flowers yellow, actinomorphic, bractless; sepals 4, free; petals 4, free; stamens 6; silicle flat, indehiscent, the single seed winged.

Isatis is a genus of about thirty Old World herbs, some of which contain substances used in dyestuffs.

Only the following species occurs in Illinois.

1. **Isatis tinctoria** L. Sp. Pl. 670. 1753. *Fig. 99.*

Annual herb from fibrous roots; stems erect, simple or branched, glabrous or pubescent near base, glaucous, to 75 cm tall; basal leaves obovate to oblong, obtuse to subacute at apex, petiolate at base, entire or coarsely toothed, to 10 cm long, glabrous; cauline leaves linear to lanceolate, obtuse to subacute at the apex, auriculate-clasping at base, entire or coarsely toothed, to 6 cm long, glabrous; corymb of racemes terminal, broad, many-flowered, the flowers 2–3 mm broad, on recurved pedicels; sepals 4, linear-lanceolate, green, glabrous, 1.0–1.5 mm long; petals 4, yellow, 2–3 mm long; silicles narrowly oblong, cuneate at base, flat, usually notched at the apex, ribbed along the sides, usually glabrous, 10–15 mm long, on pendulous pedicels, the one seed narrowly winged.

COMMON NAME: Dyer's Woad.
HABITAT: Waste ground.
RANGE: Native of Europe; rarely introduced in North America.
ILLINOIS DISTRIBUTION: Known only from Cook County.
The dyer's woad was occasionally cultivated in the past as a source for the blue dye indigo. Its collection in 1893 from Cook County represents an escape from cultivation. This plant flowers in May and June.

99. *Isatis tinctoria* (Dyer's Woad). *a.* Leaves, X½. *b.* Upper part of plant, X½.
c. Flower, X10. *d.* Sepal, X17½. *e.* Petal, X15.

100. Conringia orientalis (Hare's-ear Mustard). *a.* Habit, X½. *b, c.* Flowers,
X2½. *d.* Sepal, X6. *e.* Petal, X5. *f.* Fruit, X¾. *g.* Seed, X25.

28. *Conringia Link* – Hare's-ear Mustard

Annual, glabrous herbs; leaves simple, entire, the upper ones clasping; racemes terminal; flowers yellow, actinomorphic, bractless; sepals 4, free; petals 4, free; stamens 6; siliques linear, 4-angled, the valves 1- to 3-nerved, the seeds wingless, arranged in one row in each cell.

Seven European and Asian species make up the genus *Conringia*.

Only the following species occurs in Illinois.

1. **Conringia orientalis** (L.) Dumort. Fl. Belg. 123. 1827. *Fig. 100.*

Brassica orientalis L. Sp. Pl. 666. 1753.

Annual herb from fibrous roots; stems somewhat succulent, erect, simple or branched, glabrous, to 75 cm tall; basal leaves oblong, obtuse at the apex, cuneate to the short-petiolate base, entire, glabrous, light green, to 10 cm long, to 5 cm broad; cauline leaves oblong-lanceolate, obtuse at the apex, round-auriculate at the clasping base, entire, glabrous, light green, to 6 cm long, to 3 cm broad; racemes terminal, several-flowered, the flowers 5–8 mm broad, on slender, ascending pedicels; sepals 4, linear-lanceolate, green, glabrous, 3.5–5.0 mm long; petals 4, pale yellow, dilated above, 7–10 mm long; siliques linear, 4-angled, glabrous, to 15 cm long, terminated by the persistent, thickened style, the pedicels slender, ascending, to 1.5 cm long, the seeds oblongoid, wingless.

COMMON NAME: Hare's-ear Mustard.

HABITAT: Along roads and railroads.

RANGE: Native of Europe; adventive in much of North America.

ILLINOIS DISTRIBUTION: Scattered throughout the state.

This is one of the most easily identified species of mustards in Illinois because of its light green, obtuse, clasping cauline leaves. The pedicels seem to be too slender to lift the long siliques into an erect position.

The flowers bloom from May to August.

29. *Camelina Crantz* – False Flax

Annual herbs; leaves entire, toothed, or pinnatifid; racemes terminal and from the uppermost leaf axils; flowers yellow, actinomorphic, bractless; sepals 4, free; petals 4, free; stamens 6; silicles marginate, abruptly beaked, the valves 1-nerved, the numerous seeds wingless, arranged in two rows in each cell.

Camelina is a genus of five species native to Europe and Asia.

KEY TO THE SPECIES OF Camelina IN ILLINOIS

1. Stems and leaves glabrous or with appressed pubescence; silicles 7–9 mm long _____1. *C. sativa*
1. Stems and leaves hirsutulous; silicles 4–5 (–7) mm long _____ _____2. *C. microcarpa*

1. **Camelina sativa** (L.) Crantz, Stirp. Austr. 1:18. 1762. *Fig. 101.*

Myagrum sativum L. Sp. Pl. 641. 1753.

Annual herb from an elongated root; stems erect, branched, glabrous or appressed-pubescent, to 65 cm tall; basal leaves lanceolate, acute at the apex, petiolate at the base, entire or sparsely toothed, glabrous to appressed-pubescent, to 6 cm long; cauline leaves similar but sagittate and clasping; racemes terminal, many-flowered, the flowers 5–7 mm broad, on slender, spreading to ascending pedicels; sepals 4, green, 2–4 mm long, glabrous; petals 4, yellow, 5–7 mm long; silicles obovoid, slightly flattened, narrowly winged, 7–9 mm long, the persistent slender style 2–3 mm long, the pedicels slender, ascending, to 3 cm long, the seeds oblongoid, wingless, pale yellow-brown, 1.0–1.5 mm long.

COMMON NAME: False Flax.

HABITAT: Along railroads; in fields.

RANGE: Native of Europe; adventive in most of North America.

ILLINOIS DISTRIBUTION: Not common; primarily in the northern half of the state.

The distinction between this species and *C. microcarpa* is not always as clear-cut as the key would seem to indicate. Both the characters of stem pubescence and silicle length are occasionally overlapping.

The seeds of *C. sativa* contain an oil at one time of economic value to Europeans.

Camelina sativa, which flowers from April to August, is far less common in Illinois than *C. microcarpa*.

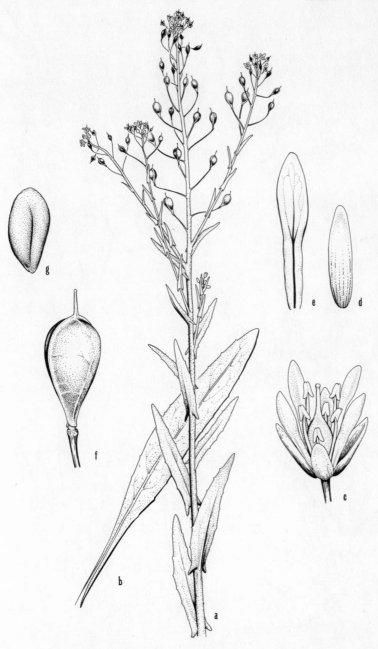

101. Camelina sativa (False Flax). *a.* Upper part of plant, X½. *b.* Basal leaf, X1½. *c.* Flower, X5. *d.* Sepal, X7½. *e.* Petal, X6. *f.* Fruit, X5. *g.* Seed, X20.

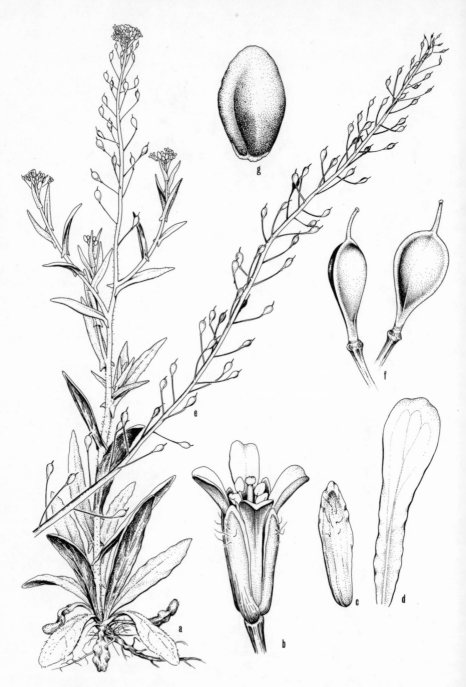

102. Camelina microcarpa (False Flax). *a.* Habit, X½. *b.* Flower, X7½. *c.* Sepal, X6. *d.* Petal, X10. *e.* Fruiting raceme, X½. *f.* Fruits, X5. *g.* Seed, X30.

2. Camelina microcarpa Andrz. in DC. Syst. 2:517. 1821. *Fig. 102.*

Annual herb from an elongated root; stems erect, simple or branched, hirsutulous, to 55 cm tall; basal leaves lanceolate, acute at the apex, cuneate to the base, entire or nearly so, hirsutulous, to 5 cm long; cauline leaves similar but sagittate and clasping; racemes terminal, many-flowered, the flowers 5–7 mm broad, on slender, spreading to ascending pedicels; sepals 4, green, 4–6 mm long, glabrous or pubescent; petals 4, yellow, 5–7 mm long; silicles obovoid, slightly flattened, broadly winged, 4–5 (–7) mm long, the persistent, slender style to 1.5 mm long, the pedicels slender, ascending, to 2.5 cm long, the seeds oblongoid, wingless, brown, about 1 mm long.

COMMON NAME: False Flax.
HABITAT: Along roads and railroads.
RANGE: Native of Europe; naturalized in much of North America.
ILLINOIS DISTRIBUTION: Occasional throughout the state. *Camelina microcarpa* is much more common in Illinois than *C. sativa*. It occurs primarily along railroads and highways.
The flowers open from April to August.

30. *Neslia Desv.* – Ball Mustard

Annual or biennial herb, with stellate pubescence; leaves entire, the upper sagittate-clasping; racemes paniculate; flowers yellow, actinomorphic, bractless; sepals 4, free; petals 4, free; stamens 6; silicles compressed, wingless, reticulate, indehiscent, slender-beaked, 1-celled, the seeds 1 (–2) per cell.

Only the following species comprises the genus.

1. Neslia paniculata (L.) Desv. Journ. Bot. 3:162. 1814. *Fig. 103.*

Myagrum paniculatum L. Sp. Pl. 641. 1753.

Annual or biennial herb from an elongated root; stems erect, simple or becoming branched above, rough-pubescent with stellate hairs, to 50 cm tall; lower leaves lanceolate, acute at the apex, cuneate to the sessile base, entire, to 6 cm long, stellate-pubescent; upper leaves linear-lanceolate, acute at the apex, sagittate and clasping at the base, to 4 cm long, stellate-pubescent; panicle terminal, the

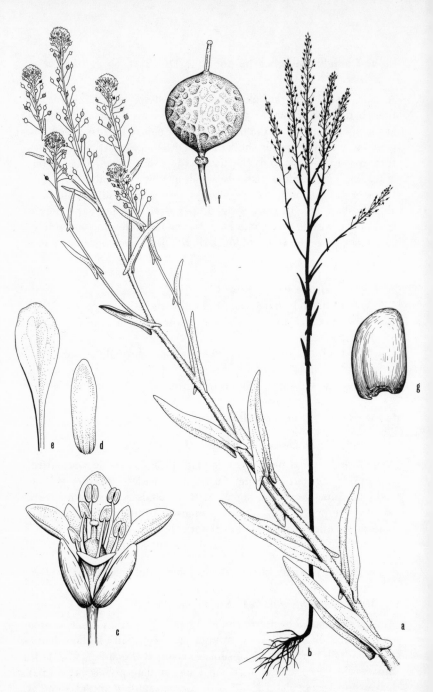

103. Neslia paniculata (Ball Mustard). *a.* Upper part of plant, X½. *b.* Habit (in silhouette), X1/6. *c.* Flower, X10. *d.* Sepal, X15. *e.* Petal, X15. *f.* Fruit, X7½. *g.* Seed, X15.

flowers about 2 mm broad, on slender, spreading pedicels; sepals 4, lanceolate, green, 1.0–1.5 mm long; petals 4, yellow, 1.5–2.5 mm long; silicles globose, flattened, reticulate, glabrous, 2–3 mm in diameter, the persistent style slender, to 1 mm long, the pedicels spreading to ascending, to nearly 1 cm long, the seed oblongoid, usually solitary.

COMMON NAME: Ball Mustard.

HABITAT: Waste ground.

RANGE: Native of Europe; uncommonly naturalized in North America.

ILLINOIS DISTRIBUTION: Known only from Du Page County.

This species is very distinctive by virtue of its globose and intricately reticulated fruit.

Ball mustard flowers during the summer.

31. *Brassica* L. – Mustard

Annual, biennial, or perennial herbs, glabrous or with unbranched hairs; basal leaves often lyrate, cauline leaves various, sometimes clasping; racemes terminal and from the uppermost leaf axils; flowers yellow (rarely white), actinomorphic, bractless; sepals 4, free; petals 4, free; stamens 6; siliques terete or angular, the valves 1- to 5-nerved, tipped by a well-developed beak, the beak sometimes 1-seeded, the silique several-seeded, unwinged, arranged in one row in each cell.

Brassica is a genus of about 80 species and numerous cultivated variants. *Brassica oleracea* is remarkable in the number of cultivated varieties which are vegetable food plants. These include var. *acephala* DC. (kale), var. *gemmifera* Zenker (brussel sprouts), var. *capitata* L. (cabbage), and var. *botrytis* L. (cauliflower and broccoli). Other cultivated members of the genus include *B. rapa* L. (rape), *B. napus* L. (turnip), *B. napobrassica* Mill. (rutabaga), and *B. kaber* (DC.) Wheeler (charlock).

KEY TO THE SPECIES OF Brassica IN ILLINOIS

1. None of the leaves auriculate and clasping at the base _____ 2
1. Some of the cauline leaves auriculate and clasping at the base _____ 5
   2. Leaves all pinnatifid; fruits densely white-bristly _____ 1. *B. hirta*
   2. At least the uppermost leaves merely toothed, not pinnatifid; fruits glabrous or nearly so _____ 3
3. Plants more or less hispid or hirsute, green; fruiting pedicels 3–7 mm long _____4

3. Plants glabrous or nearly so, glaucous; fruiting pedicels 7–10 mm long
--- 4. *B. juncea*

   4. Fruits 2.5–5.5 cm long, ascending, the 1-seeded beak 10–15 mm
    long; upper leaves sessile ------------------------- 2. *B. kaber*

   4. Fruits 1–2 cm long, erect, the seedless beak 1–3 mm long; upper
    leaves short-petiolate ---------------------------- 3. *B. nigra*

5. Plants glaucous --- 6

5. Plants green -- 7. *B. napus*

   6. Leaves not thick; petals 7–10 mm long; young leaves setose -------
    --- 5. *B. rapa*

   6. Leaves thick; petals 15 mm long or longer; young leaves glabrous
    -- 6. *B. oleracea*

1. Brassica hirta Moench, Meth. Pl. Suppl. 84. 1802. *Fig. 104.*
Synapis alba L. Sp. Pl. 668. 1753.
Brassica alba (L.) Rabenh. Fl. Lusatica 1:184. 1839, non Gilib.
(1782).

Coarse annual herbs from thickened roots; stems erect, branched,
hispid, to 1 m tall; lower leaves obovate in outline, pinnatifid, to
20 cm long, the terminal lobe larger than the other lobes, dentate,
more or less hispidulous; upper leaves oblong to lanceolate, pinnatifid, petiolate, hispidulous, to 10 cm long; racemes terminal, several-flowered, the flowers 1.0–1.5 cm broad, on rather stout, ascending pedicels; sepals 4, linear, green, more or less pubescent,
4–7 mm long; petals 4, yellow, 1.0–1.5 cm long; siliques to 3.5 cm
long (including the beak), terete, constricted between the seeds,
white-hispid, the beak often as long as the body of the silique, flat,
straight or arcuate, the seeds globose, light brown or yellow.

COMMON NAME: White Mustard.
HABITAT: Along railroads; fields.
RANGE: Native of Europe and Asia; introduced in various
parts of the United States.
ILLINOIS DISTRIBUTION: Scattered throughout the state.
If one follows McVaugh (1949) in rejecting all of Gilibert's
new binomials because of Gilibert's lack of consistency in
using binomial nomenclature, then the correct name for
this species would be *Brassica alba* (L.) Rabenh.
   The white mustard is grown as a source for mustard (from
the seeds) and for salads (from the young leaves). Several horticul-
tural forms have been developed.
   This species flowers from April to July.

104. Brassica hirta (White Mustard). *a.* Upper part of plant, X½. *b.* Flower, X2½. *c.* Sepal, X7½. *d.* Petal, X4. *e.* Fruit, X1½. *f.* Seed, X10.

105. *Brassica kaber* (Charlock). *a.* Upper part of plant, X½. *b.* Lower leaf, X½. *c.* Flower, X2½. *d.* Sepal, X4. *e.* Petal, X3. *f.* Fruit, X1¾. *g.* Seed, X10.

2. Brassica kaber (DC.) Wheeler, Rhodora 40:306. 1938.
Sinapis kaber DC. Veg. Syst. 2:617. 1821.

Annual herbs from elongated roots; stems erect, branched, hispid or becoming more or less glabrous, to 75 cm tall; lower leaves obovate in outline, mostly lyrate, to 20 cm long, the terminal lobe larger than the other lobes, dentate, more or less pilose; upper leaves ovate to oblong-ovate, obtuse to subacute at the apex, sessile at the base, sometimes lobed, dentate, to 8 cm long, pilose; racemes terminal, several-flowered, the flowers 1.2–2.0 cm broad, on slender to stout, spreading to ascending pedicels; sepals 4, linear, green, glabrous or pubescent, often reflexed, 4–8 mm long; petals 4, yellow, long-clawed, 1.0–1.7 cm long; siliques to 4.5 cm long (including the beak), terete, constricted between the seeds, more or less glabrous, the beak shorter than the body of the silique, flat, usually curved, the pedicels to 7 mm long, the seeds globose, brown.

The typical variety of this species is not recorded from North America, although two other varieties are naturalized on much of the continent.

KEY TO THE VARIETIES OF Brassica kaber IN ILLINOIS

1. Siliques slightly torulose, 3–4 mm thick _____
_____2a. *B. kaber* var. *pinnatifida*
1. Siliques strongly torulose, 1.5–2.0 mm thick _____
_____2b. *B. kaber* var. *schkuhriana*

2a. Brassica kaber (DC.) Wheeler var. **pinnatifida** (Stokes) Wheeler, Rhodora 40:308. 1938. *Fig. 105.*
Sinapis arvensis L. β *pinnatifida* Stokes, Bot. Mater. Medic. 3:478. 1812.
Brassica arvensis Rabenh. Fl. Lusatica 1:184. 1839, non L. (1753).
Brassica sinapistrum Boiss. Voy. Esp. 2:39. 1839–45.

Siliques slightly torulose, 3–4 mm thick.

COMMON NAMES: Charlock; Crunchweed.

HABITAT: Fields; along roads; along railroads.

RANGE: Native of Europe and Asia; naturalized through-out North America.

ILLINOIS DISTRIBUTION: Occasional throughout the state. The charlock is one of the more commonly found weedy species of *Brassica* in Illinois.

Wheeler (1938) considers this taxon to be a variety of the Eurasian *B. kaber*. It has no significant economic impor-tance.

This variety flowers from April to July.

2b. Brassica kaber (DC.) Wheeler var. **schkuhriana** (Reichenb.) Wheeler, Rhodora 40:308. 1938.

Sinapis schkuhriana Reichenb. Icon. Fl. Germ. 2:20, t. 87. 1837–38.

Siliques strongly torulose, 1.5–2.0 mm thick.

COMMON NAME: Field Mustard.

HABITAT: Waste ground.

RANGE: Native of Europe and Asia; naturalized through-out North America, although much less common than var. *pinnatifidum*.

ILLINOIS DISTRIBUTION: Rare and scattered throughout the state.

I am following the work of Wheeler in segregating var. *schkuhriana* from both var. *pinnatifida* and var. *kaber*.

3. Brassica nigra (L.) Koch in Roehl, Deutsche Fl. ed. 3, 4: 713. 1833. *Fig. 106.*

Sinapis nigra L. Sp. Pl. 668. 1753.

Annual herb from elongated roots; stems erect, branched, hispid to nearly glabrous, to 3 m tall; lower leaves deeply pinnatifid, petio-late, the terminal lobe much larger than the other lobes, dentate, more or less hispid on both surfaces; middle cauline leaves pinnat-ifid or merely similar to the basal but smaller; uppermost cauline leaves oblong to lanceolate, entire; racemes terminal and from the uppermost leaf axils, several-flowered, the flowers 6–12 mm broad, on slender, erect, appressed pedicels; sepals 4, lanceolate, green, 3–6 mm long; petals 4, bright yellow, 6–12 mm long; siliques lin-

106. Brassica nigra (Black Mustard). *a.* Upper part of plant, X¼. *b.* Habit (in silhouette), X1/12. *c.* Flower, X3. *d.* Sepal, X7½. *e.* Petal, X5. *f.* Flowering and fruiting raceme, X½. *g.* Fruit, X5. *h.* Seed, X12½.

107. *Brassica juncea* (Indian Mustard). *a.* Habit X¼. *b.* Flower, X2½. *c.* Sepal, X6. *d.* Petal, X4. *e.* Fruit, X1½. *f.* Seed, X10.

ear, 1–2 cm long (including the beak), 4-angled, not constricted between the seeds, more or less glabrous, abruptly narrowed to the slender 1–3 mm long beak, the beak without a seed, the pedicels erect, appressed, to 5 mm long, the seeds dark brown or black, minute.

COMMON NAME: Black Mustard.
HABITAT: Fields, along roads.
RANGE: Native of Europe and Asia; naturalized throughout North America.
ILLINOIS DISTRIBUTION: Occasional throughout the state. Black mustard is a coarse weed of cultivated fields. Robust specimens may attain a height of three meters.
Brassica nigra differs from *B. kaber* by its shorter, erect-appressed siliques with a short unseeded beak.
The flowers open from April to October.

4. **Brassica juncea** (L.) Coss, Bull. Soc. Bot. Fr. 6:609. 1859. *Fig. 107.*
Sinapis juncea L. Sp. Pl. 668. 1753.

Annual herb from elongated roots; stems erect, branched, glabrous or occasionally setose when young, glaucous, to 1.3 m tall; basal leaves lyrate to runcinate-pinnatifid, oval to obovate in outline, obtuse at the apex, petiolate at the base, glabrous or setose when young, to 30 cm long; lower cauline leaves similar but smaller; upper cauline leaves lanceolate to linear, entire to dentate, glabrous, sessile or nearly so; racemes terminal and from the uppermost leaf axils, several-flowered, the flowers 1.2–2.0 cm broad, the pedicels strongly ascending to erect; sepals 4, linear-lanceolate, green, glabrous, 4–8 mm long; petals 4, bright yellow, 8–15 mm long; siliques linear, to 6 cm long (including the beak), not constricted between the seeds, glabrous, the slender beak 1–2 cm long, without a seed, the pedicels ascending to erect, to 10 mm long, the seeds brown.

COMMON NAMES: Indian Mustard; Leaf Mustard; Chinese Mustard.

HABITAT: Fields, along roads and railroads.

RANGE: Native of Europe and Asia; naturalized throughout North America.

ILLINOIS DISTRIBUTION: Occasional throughout the state. This mustard is sometimes cultivated as a source of vegetable greens. It readily escapes from cultivation.

Brassica juncea is conspicuously glaucous, thereby easily distinguishing it from either *B. kaber* or *B. nigra*.

The flowers open from April to September.

5. **Brassica rapa** L. Sp. Pl. 666. 1753. *Fig. 108.*
Brassica campestris L. Sp. Pl. 666. 1753.

Biennial herb from a nontuberous, much-branched root; stems erect, branched, glabrous or nearly so, glaucous, to 1 m tall; basal leaves lyrate-pinnatifid, to 30 cm long, petiolate, setose or glabrous; lower cauline leaves similar to the basal leaves but smaller; upper cauline leaves oblong to lanceolate, obtuse to acute at the apex, auriculate and clasping at the base, dentate, glabrous; racemes terminal and from the uppermost leaf axils, several-flowered, the flowers 8–12 mm broad, the pedicels spreading to ascending; sepals 4, linear-lanceolate, 4–6 mm long; petals 4, bright yellow, 7–10 mm long; siliques linear, to 8 cm long (including the beak), glabrous, not constricted between the seeds, the slender beak to 2 cm long, without a seed, the pedicels spreading to ascending, the seeds dark brown.

COMMON NAMES: Field Mustard; Bird's Rape.

HABITAT: Fields, waste ground.

RANGE: Native of Europe and Asia; naturalized throughout North America.

ILLINOIS DISTRIBUTION: Scattered in most regions of Illinois.

The field mustard is an occasionally encountered weed in fields and waste places.

The young stems and leaves can be cooked and eaten as greens.

The glaucous stems and leaves distinguish this species from *B. napus*. It differs from *B. oleracea*, another glaucous species, by its setose young leaves and smaller flowers.

The flowers bloom from April to September.

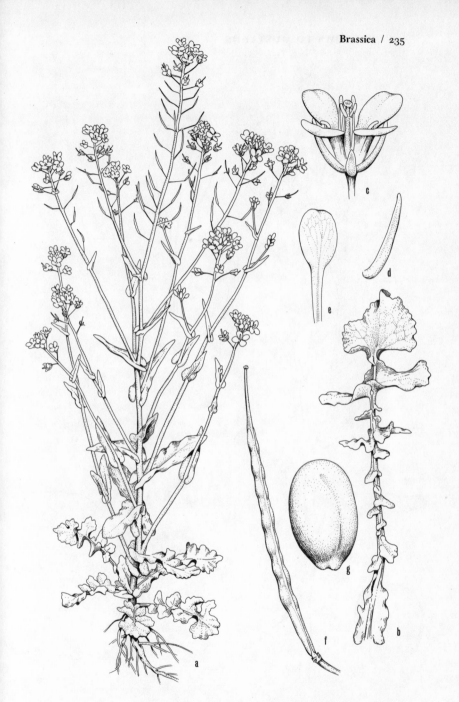

108. *Brassica rapa* (Field Mustard). *a*. Habit, X⅛. *b*. Lower leaf, X½. *c*. Flower, X3. *d*. Sepal, X5. *e*. Petal, X4. *f*. Fruit, X1. *g*. Seed, X20.

109. Brassica oleracea (Mustard). *a.* Upper part of plant, X½. *b.* Flower, X2½.
c. Sepal, X4. *d.* Petal, X2½. *e.* Fruit, X1. *f.* Seed, X12½.

6. Brassica oleracea L. Sp. Pl. 667. 1753. *Fig. 109.*

Annual or biennial herb from thickened roots; stems erect, branched, glabrous, glaucous, to 1 m tall; basal and lower cauline leaves oblong or obovate in outline, pinnatifid or dentate, fleshy, glabrous, glaucous, to 45 cm long, petiolate; upper cauline leaves oblong to lanceolate, obtuse to acute at the apex, auriculate and clasping at the base, fleshy, glabrous, glaucous; racemes terminal and from the uppermost leaf axils, several-flowered, the flowers 1.5–2.5 cm broad, the pedicels spreading to ascending; sepals 4, lanceolate, green, glabrous, 6–12 mm long; petals 4, yellow-white, 1.5–2.5 cm long; siliques linear, to 10 cm long (including the beak), glabrous, not constricted between the seeds, glabrous, the slender beak to 2 cm long, without a seed, the pedicels spreading to ascending, the seeds brown.

COMMON NAME: Mustard.

HABITAT: Fields.

RANGE: Native of Europe and Asia; rarely adventive in North America.

ILLINOIS DISTRIBUTION: Known only from Peoria County (along Galena Road, May 11, 1950, *V. H. Chase 10977*).

Brassica oleracea has many cultivated varieties which are important food crops.

The most commonly cultivated ones in Illinois are the cabbage (var. *capitata* L.) and cauliflower and broccoli (var. *botrytis* L.).

The flowers open from April to September.

7. Brassica napus L. Sp. Pl. 666. 1753. *Figs. 110, 111.*

Biennial herb from a thick, reddish, white-fleshed tuberous root; stems erect, branched, glabrous or nearly so, green, to 1 m tall; basal leaves pinnatifid, to 45 cm long, petiolate, setose; cauline leaves pinnatifid to dentate to entire, to 25 cm long, setose, auriculate and clasping at the base; racemes terminal and from the uppermost leaf axils, the flowers 7–10 mm broad, the pedicels spreading to ascending; sepals 4, linear-lanceolate, green, glabrous, 3–5 mm long; petals 4, bright yellow, 7–10 mm long; siliques linear, to 6 cm long (including the beak), glabrous, not constricted between the seeds, the slender beak to 1.5 cm long, without a seed, the pedicels spreading or ascending, the seeds dark brown.

110. Brassica napus (Turnip).   *a.* Habit (autumnal stage), X1/16.   *b.* Basal leaf, X½.
c. Flower, X3.   *d.* Sepal, X6.   *e.* Petal, X5.

111. Brassica napus (Turnip). *f.* Upper part of plant, X1/6. *g.* Root and lowest leaf, X1/6. *h.* Fruit, X2½. *i.* Seed, X12½.

COMMON NAME: Turnip.

HABITAT: Cultivated ground.

RANGE: Native of Europe and Asia; escaped from cultivation in some parts of North America.

ILLINOIS DISTRIBUTION: Known from several Illinois counties.

The turnip is a common garden plant which rarely escapes from or persists after cultivation.

Some botanists have called this species *B. rapa*, using the binomial *B. napus* for the species recognized in this work as *B. rapa*.

Brassica napus flowers from May to September.

32. *Erucastrum Presl* – Dog Mustard

Annual or biennial herbs; leaves pinnatifid to bipinnatifid; inflorescence racemose, the lower flowers bracteate; flowers yellow, actinomorphic; sepals 4, free; petals 4, free; stamens 6; siliques linear, 4-angled, torulose, beaked, the valves keeled and delicately nerved, the seeds arranged in one row.

Erucastrum is a small genus native primarily to the Mediterranean region. This genus differs from *Brassica* by its bracteate lower flowers.

Only the following species occurs in Illinois.

> **1. Erucastrum gallicum** (Willd.) O. E. Schulz, Bot. Jahrb. 54:
> Beibl. 119:56. 1916. *Fig. 112.*
> *Sisymbrium gallicum* Willd. Enum. Pl. Nort. Berol. 678. 1809.

Annual herb; stems erect, retrorsely pubescent, to 75 cm tall; leaves oblong in outline, pinnatifid to bipinnatifid, to 20 cm long, pubescent, petiolate; racemes terminal and from the uppermost leaf axils, several-flowered, the flowers 4–7 mm broad, on spreading to ascending pedicels, the lower flowers bracteate; sepals 4, lanceolate, pubescent, 2–3 mm long; petals 4, pale yellow, clawed, 4–5 mm long; siliques linear, torulose, glabrous, to 3.5 cm long (including the 3–6 mm long conical beak), the pedicels spreading to ascending, the small seeds ovoid.

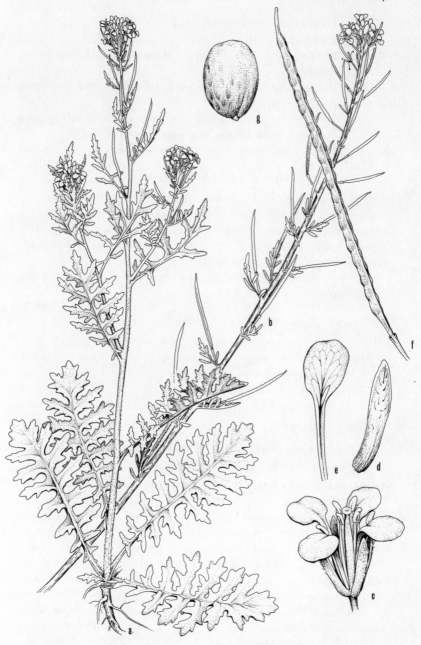

112. *Erucastrum gallicum* (Dog Mustard). *a.* Habit, X⅓. *b.* Flowering and fruiting raceme, X⅓. *c.* Flower, X5. *d.* Sepal, X10. *e.* Petal, X7½. *f.* Fruit, X2½. *g.* Seed, X12½.

COMMON NAME: Dog Mustard.

HABITAT: Fields; along roads; along railroads.

RANGE: Native of Europe; adventive in eastern North America.

ILLINOIS DISTRIBUTION: Scattered throughout the state, but not common.

Erucastrum gallicum is unique among the Illinois species of Brassicaceae by the bracteate lower flowers.

The flowers appear from May to September.

33. *Sisymbrium* L. – Hedge Mustard

Annual or biennial herbs; leaves pinnatifid to toothed to entire, basal and cauline; inflorescence racemose, the flowers yellow, actinomorphic, bractless; sepals 4, free; petals 4, free; stamens 6; siliques linear, cylindric, short-beaked, the valves 1- to 3-nerved, the seeds wingless, arranged in one row.

Sisymbrium is a genus of about a dozen species native to the Old World.

KEY TO THE SPECIES OF *Sisymbrium* IN ILLINOIS

1. Petals up to 3 mm long; siliques up to 2 cm long _____ 1. *S. officinale*
1. Petals 5–6 mm long; siliques 2–10 cm long _____ 2
2. At least the upper leaves divided into threadlike divisions; siliques 5–10 cm long ----------------------------------- 2. *S. altissimum*
2. Leaves divided into triangular lobes; siliques 2–4 cm long -----------
--- 3. *S. loeselii*

1. Sisymbrium officinale (L.) Scop. Fl. Carn., ed. 2, 2:26. 1772.
Erysimum officinale L. Sp. Pl. 660. 1753.

Annual herb; stems erect, branched, hirsute at the base, becoming less pubescent above, to 1 m tall; basal leaves runcinate-pinnatifid, 5- to 13-lobed, the lobes dentate, pubescent, petiolate; cauline leaves becoming less lobed to dentate or entire, acute to obtuse, shorter petiolate, sparsely pubescent to glabrous or nearly so; racemes terminal, several-flowered, the flowers 3–4 mm broad, on erect-appressed pedicels, bractless; sepals 4, lanceolate, green, more or less pubescent, 1.0–1.5 mm long; petals 4, yellow, 2.5–3.0 mm long; siliques broadest at base, uniformly tapering to a subulate beak, glabrous or pubescent, to 2 cm long, the pedicels erect-appressed, to 2 mm long, the seeds wingless.

The short, conical siliques distinguish this species from other members of the genus in Illinois.

Two varieties occur in Illinois.

1. Siliques pubescent _____ 1a. *S. officinale* var. *officinale*
1. Siliques glabrous or nearly so _____ 1b. *S. officinale* var. *leiocarpum*

1a. Sisymbrium officinale (L.) Scop. var. **officinale** *Fig. 113a–h.*
Siliques pubescent.

COMMON NAME: Hedge Mustard.
HABITAT: Fields; along roads; along railroads; barnyards.
RANGE: Native of Europe and Asia; naturalized through-out North America.
ILLINOIS DISTRIBUTION: Scattered in Illinois, but much less common than var. *leiocarpum*.
The siliques are uniformly short-pubescent throughout in this variety.
Time of flowering is May to October.

1b. Sisymbrium officinale (L.) Scop. var. **leiocarpum** DC.
Prodr. 1:191. 1824. *Fig. 113i.*
Siliques glabrous or nearly so.

COMMON NAME: Hedge Mustard.
HABITAT: Fields; barnyards; along roads; along railroads.
RANGE: Native of Europe and Asia.
ILLINOIS DISTRIBUTION: Occasional throughout the state.
This variety, in which the siliques are glabrous or only sparsely pubescent, is far more common in Illinois than var. *officinale*. It is a common inhabitant of barnyards.
The flowers bloom from May to October.

2. Sisymbrium altissimum L. Sp. Pl. 659. 1753. *Figs. 114, 115.*
Norta altissima (L.) Britt. in Britt. & Brown, Illus. Fl. ed. 2,
2:174. 1913.
Annual or biennial herb; stems erect, branched, sparsely hirsute at the base, more or less glabrous above, to 1.3 m tall; basal leaves runcinate-pinnatifid, 5- to 19-lobed, the lobes dentate, sparsely pubescent to glabrous, petiolate; upper cauline leaves divided into threadlike divisions, glabrous or nearly so; racemes terminal and from the uppermost leaf axils, several-flowered, the flowers 6–8 mm broad, the pedicels spreading to ascending; sepals 4, linear-

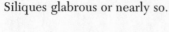

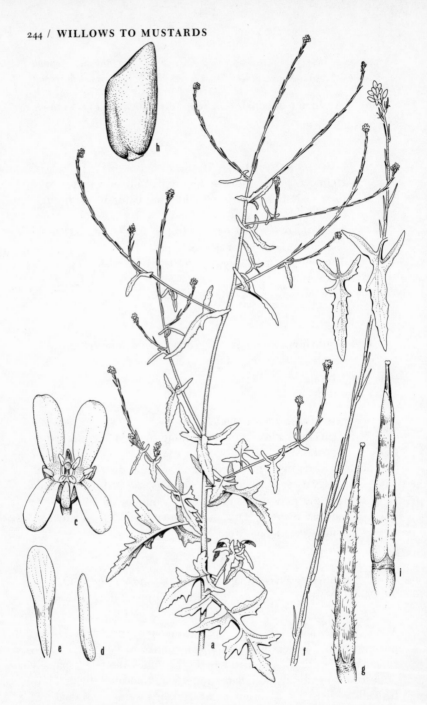

113. Sisymbrium officinale (Hedge Mustard). *a.* Upper part of plant, X¼. *b.* Upper leaves, X¾. *c.* Flower, X5. *d.* Sepal, X15. *e.* Petal, X10. *f.* Flowering and fruiting raceme, X½. *g.* Fruit, X3. *h.* Seed, X25. var. *leiocarpum. i.* Fruit, X3.

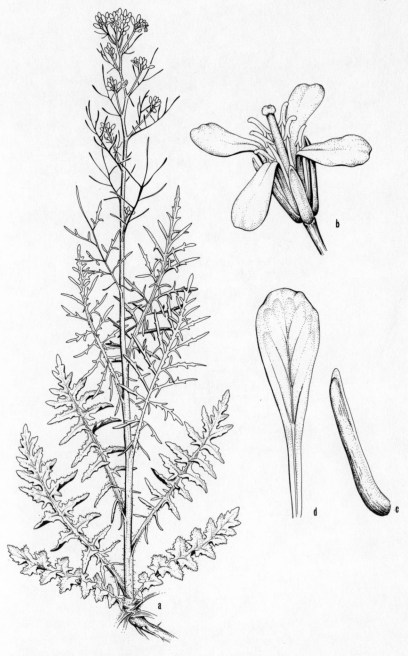

114. *Sisymbrium altissimum* (Tumble Mustard). *a.* Habit, X1/6. *b.* Flower, X6.
c. Sepal, X15. *d.* Petal, X9.

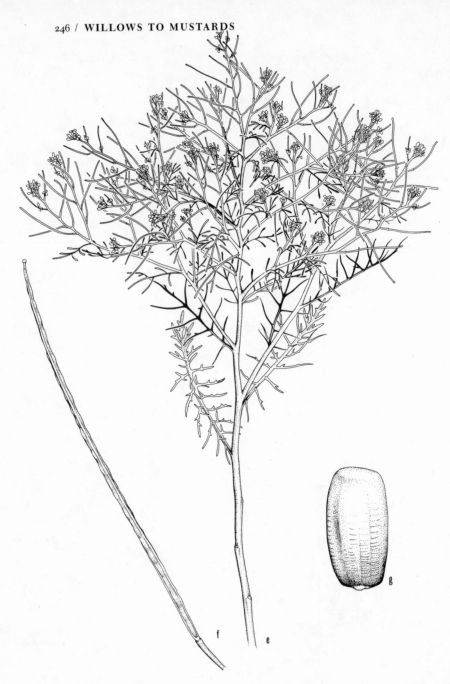

115. Sisymbrium altissimum (Tumble Mustard). *e.* Upper part of plant, X1/6.
f. Fruit, X1. *g.* Seed, X32½.

lanceolate, green, glabrous, 2–3 mm long; petals 4, pale yellow,
6–8 mm long; siliques narrowly linear, cylindric, rigid, glabrous,
5–10 cm long, short-beaked, the pedicels spreading to ascending,
to nearly 1 cm long, the seeds oblongoid, unwinged.

COMMON NAME: Tumble Mustard.
HABITAT: Fields; along roads; along railroads.
RANGE: Native of Europe; naturalized throughout North
America.
ILLINOIS DISTRIBUTION: Occasional throughout the state.
Sisymbrium altissimum differs from other species of *Sisym-
brium* in Illinois by its longer siliques and threadlike di-
visions of the upper leaves.
The flowers appear from May to August.

3. **Sisymbrium loeselii** L. Cent. Pl. 1:18. 1755. *Fig. 116.*

Annual herb; stems erect, branched, retrorsely hispid, to 1 m tall;
basal leaves runcinate-pinnatifid, 5- to 13-lobed, the lobes triangular, dentate, sparsely pubescent to glabrous, petiolate; upper leaves
similar but smaller and less lobed; racemes terminal, several-flowered, the flowers 6–8 mm broad, the pedicels threadlike, spreading
to ascending; sepals 4, linear-lanceolate, green, glabrous or nearly
so, 2–3 mm long; petals 4, yellow, 5–6 mm long; siliques narrowly
linear, cylindric, glabrous or pubescent, 2–4 cm long, short-
beaked, the pedicels spreading to ascending, to 1.5 cm long, the
seeds oblongoid, unwinged.

COMMON NAME: Tall Hedge Mustard.
HABITAT: Barnyards; along railroads.
RANGE: Native of Europe; adventive in North America.
ILLINOIS DISTRIBUTION: Occasional in the extreme north-
ern counties, rare elsewhere.
Swink (1974) reports this species increasing in abundance
in the northeastern corner of Illinois, where it occurs pri-
marily in barnyards.
Sisymbrium loeselii differs from *S. altissimum* by its short-
er siliques.
The flowers are borne from May to October.

34. *Rorippa* Scop. – Yellow Cress

Annual, biennial, or perennial herbs; leaves pinnately compound,
pinnatifid, or simple and unlobed, basal and cauline; inflorescence

116. *Sisymbrium loeselii* (Tall Hedge Mustard). *a.* Upper part of plant, X1/6. *b.* Lower leaf, X½. *c.* Flower, X5. *d.* Sepal, X12½. *e.* Petal, X7½. *f.* Fruit, X3½. *g.* Seed, X27½.

racemose, the flowers yellow, actinomorphic, bractless; sepals 4, free; petals 4, free; stamens usually 6; silicles or siliques terete, the valves nerveless, the seeds turgid, numerous, wingless, arranged in two rows.

There are about fifty species of *Rorippa* native primarily to the northern hemisphere.

KEY TO THE SPECIES OF Rorippa OF ILLINOIS

1. Petals over 4 mm long, longer than the sepals _____ 2
1. Petals absent, or up to 2 mm long, never longer than the sepals _____ 3
  2. Leaves without auricles at base; seeds up to 0.8 mm long _____
  _____ 1. *R. sylvestris*
  2. Leaves with auricles at base; seeds about 1 mm long _____
  _____2. *R. sinuata*
3. Petals up to 0.5 mm long; stamens 4; fruits 6–10 times longer than the pedicels; style nearly absent; seeds 175 or more per fruit _____
  _____3. *R. sessiliflora*
3. Petals 1–2 mm long; stamens 6; fruits at most only 4 times longer than the pedicels; style present; seeds less than 75 per fruit _____ 4
  4. Petals 1.0–1.2 mm long; fruits 2–4 times longer than the pedicels
  _____4. *R. truncata*
  4. Petals 1.7–2.0 mm long; fruits shorter than to up to 2 times longer than the pedicels _____ 5. *R. islandica*

1. **Rorippa sylvestris** (L.) Bess. Enum. 27. 1821. *Fig. 117.*
Sisymbrium sylvestre L. Sp. Pl. 657. 1753.
Nasturtium sylvestre (L.) R. Br. in Ait. Hort. Kew., ed. 2, 4: 110. 1812.
Radicula sylvestris (L.) Druce, List Brit. Plants 4. 1908.

Perennial from rhizomes; stems ascending, branched, glabrous, to 50 cm tall; leaves pinnately compound into 5–11 leaflets, petiolate, glabrous, to 10 cm long, the leaflets oblong to lanceolate, dentate; racemes terminal and from the uppermost leaf axils, several-flowered, the flowers 6–9 mm broad, on spreading to ascending pedicels; sepals 4, lanceolate, green, glabrous, 2.5–4.0 mm long; petals 4, yellow, 6–8 mm long; siliques linear, cylindric, glabrous, to 2.5 cm long, the style persistent as a slender beak to 1 mm long, the pedicels spreading to ascending, about as long as the flower, the seeds up to 0.8 mm long.

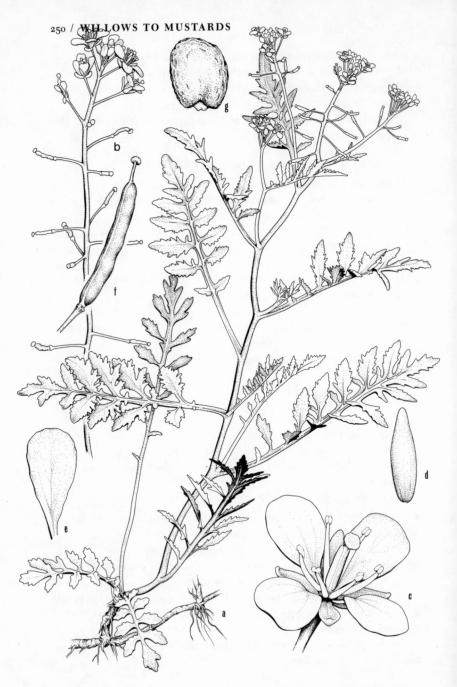

117. Rorippa sylvestris (Creeping Yellow Cress). *a.* Habit, X½. *b.* Flowering and fruiting raceme, X1. *c.* Flower, X4. *d.* Sepal, X7. *e.* Petal, X4. *f.* Fruit, X2. *g.* Seed, X25.

COMMON NAME: Creeping Yellow Cress.

HABITAT: Moist disturbed areas.

RANGE: Native of Europe and Asia; naturalized in eastern North America.

ILLINOIS DISTRIBUTION: Occasional throughout the state.

Rorippa sylvestris and *R. sinuata* are the only rhizomatous species of *Rorippa* in Illinois. *Rorippa sylvestris* sends up aerial branches from all along the rhizome, while *R. sinuata* has stems clustered from a definite basal rosette. In addition, none of the pinnately compound leaves of *R. sylvestris* are auriculate at the base, while some of the pinnatifid leaves of *R. sinuata* are auriculate.

The flowers bloom from May to September.

2. **Rorippa sinuata** (Nutt.) Hitchc. Spring Fl. Manhattan, Kans. 18. 1894. *Fig. 118.*

Nasturtium sinuatum Nutt. ex Torr. & Gray, Fl. N. Am. 1:73. 1838.

Radicula sinuata (Nutt.) Greene, Leaflets 1:113. 1905.

Perennial from rhizomes; stems clustered from a basal rosette, ascending, branched, glabrous, to 35 cm tall; leaves oblong to broadly lanceolate in outline, pinnatifid, to 7.5 cm long, the lobes linear to oblong, obtuse, entire or sparsely dentate, glabrous, at least the middle and upper cauline leaves auriculate at the base; racemes terminal and from the uppermost leaf axils, several-flowered, the flowers 5–7 mm broad, on ascending pedicels; sepals 4, linear-lanceolate, green, glabrous, 2.5–4.0 mm long; petals 4, yellow, 6–8 mm long; siliques linear, cylindric, falcate, glabrous, to 1.5 cm long, the style persistent as a slender beak to 3 mm long, the pedicels ascending, to 1 cm long, the seeds about 1 mm long.

COMMON NAME: Spreading Yellow Cress.

HABITAT: Stream banks.

RANGE: Ontario to Washington, south to California, Texas, and Illinois.

ILLINOIS DISTRIBUTION: Throughout the state, but not common.

Rorippa sinuata and *R. sylvestris* are similar in possessing rhizomes and in having the petals longer than the sepals. *Rorippa sinuata* may be distinguished from *R. sylvestris* by its pinnatifid instead of pinnately compound leaves, its

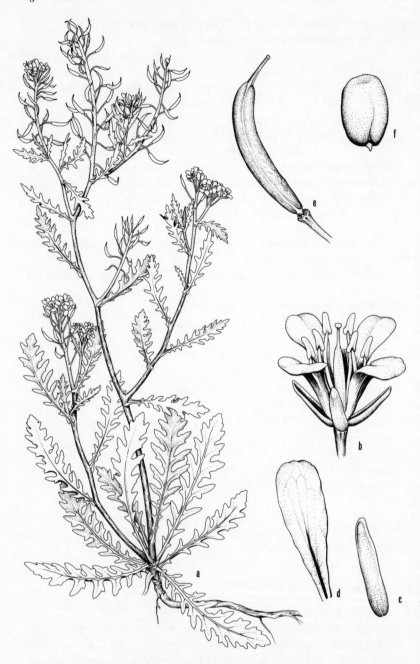

118. *Rorippa sinuata* (Spreading Yellow Cress). *a*. Habit, X½. *b*. Flower, X4.
c. Sepal, X7. *d*. Petal, X5. *e*. Fruit, X2½. *f*. Seed, X20.

auriculate-based upper leaves, and by its slightly larger seeds.
This species flowers from April to July.

3. **Rorippa sessiliflora** (Nutt.) Hitchc. Spring Fl. Manhattan,
Kans. 18. 1894. *Fig. 119.*

Nasturtium sessiliflorum Nutt. ex Torr. & Gray, Fl. N. Am.
1:73. 1838.

Radicula sessiliflora (Nutt.) Greene, Leaflets 1:113. 1905.

Annual or biennial from thickened roots; stems erect, simple or
branched, glabrous, to 35 cm tall; basal leaves oblong in outline, to
10 cm long, petiolate, lyrate-pinnatifid, glabrous; lower cauline
leaves similar to the basal leaves; upper cauline leaves oblong to
obovate, crenate, short-petiolate to nearly sessile, glabrous; racemes terminal and from the uppermost leaf axils, several-flowered,
the flowers 1.5–2.0 mm broad, sessile or on pedicels to 1 mm long;
sepals 4, lanceolate, green, glabrous, 0.5–1.0 mm long; petals 4,
yellow, to 0.5 mm long or even absent; stamens 4; siliques narrowly
oblongoid to oval, glabrous, 6–12 mm long, minutely beaked, on
ascending pedicels to 1 mm long, the seeds minute, at least 175
per silique.

COMMON NAME: Sessile-flowered Cress.

HABITAT: Along rivers and streams.

RANGE: Northwestern Indiana to Wisconsin and Minnesota, south to Texas and Florida; Nebraska; Virginia.

ILLINOIS DISTRIBUTION: Occasional to common in the
southern four-fifths of the state, less common elsewhere.

This species is readily recognized by its nearly sessile flowers and fruits. It is also the only *Rorippa* in Illinois with
four stamens.

This is one of the most common species of flowering plants
along river banks in the southern half of Illinois.

Rorippa sessiliflora flowers from April to November.

4. **Rorippa truncata** (Jepson) Stuckey, Sida 2:414. 1966. *Fig.
120.*

Radicula sinuata (Nutt.) Greene var. *truncata* Jepson, Fl. Pl.
Cal. 424. 1925.

Annual or biennial herb from thickened roots; stems ascending to
erect, branched, glabrous, to 35 cm tall; leaves pinnately compound
to pinnatifid, oblong in outline, to 10 cm long, the segments obtuse,
entire to repand to dentate, petiolate, glabrous; racemes terminal

and from the uppermost leaf axils, several-flowered, the flowers to 2 mm broad, on ascending pedicels 1–3 mm long; sepals 4, lanceo-late, green, glabrous, 1.0–1.5 mm long; petals 4, yellow, 1.0–1.2 mm long; stamens 6; siliques obtuse to truncate at the apex, nar-rowly oblongoid to oval, glabrous, to (2–) 3–5 (–8) mm long, mi-nutely and slenderly beaked, on ascending pedicels 1–3 mm long, the seeds up to 75 per silique.

COMMON NAME: Blunt-leaved Yellow Cress.

HABITAT: Along rivers.

RANGE: West Virginia to Michigan to British Columbia, south to California, Texas, and southwestern Illinois.

ILLINOIS DISTRIBUTION: Known first from St. Clair Coun-ty (East St. Louis, June 11, 1895, *Letterman s.n.*); more recently collected in Cass and Jackson counties.

This very rare Illinois species is similar in appearance to *R. sessiliflora* from which it differs by its longer pedicels and fewer seeds per silique. It differs from *R. islandica* by its shorter petals and shorter pedicels.

The first collection known from Illinois came from the banks of the Mississippi River in St. Clair County.

Stuckey (1966) has presented evidence to show that our species, *R. obtusa*, should actually be called *R. truncata*. *Rorippa obtusa*, which Stuckey says should be called *R. teres* (Michx.) Stuckey, does not occur in Illinois.

This species flowers from May to September.

5. **Rorippa islandica** (Oeder) Borbas, Bal. Tav. Partm. 392. 1900.

Sisymbrium islandicum Oeder, Fl. Dan. 3(7):8, t. 409. 1768.

Annual or biennial herbs from thickened roots; stems erect, simple or branched, glabrous or hirsutulous, to 1.2 m tall; leaves oblong or oblanceolate in outline, pinnately compound, pinnatifid, or only toothed, firm or membranaceous, glabrous to hirsutulous, the lower leaves usually petiolate, the middle leaves usually auriculate and clasping, the upper leaves usually sessile or nearly so; racemes ter-minal and from the uppermost leaf axils, several-flowered, the flow-ers 2–4 mm broad, on spreading or curved-ascending pedicels 3 mm long or longer; sepals 4, lanceolate, green, glabrous or pubes-cent, 1.5–2.0 mm long; petals 4, oblanceolate, yellow, 1.5–2.0 mm long; stamens 6; siliques ellipsoid to subglobose, sometimes curved, glabrous, 2–10 mm long, short-beaked, on spreading or curved-

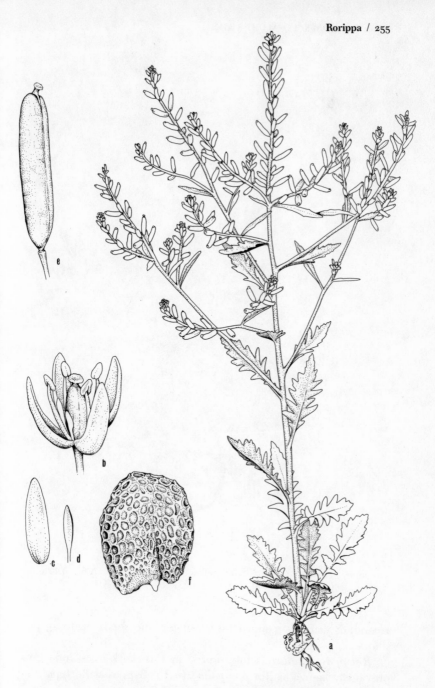

119. Rorippa sessiliflora (Sessile-flowered Cress). *a.* Habit, X½. *b.* Flower, X20.
c. Sepal, X22½. *d.* Petal, X22½. *e.* Fruit, X4. *f.* Seed, X60.

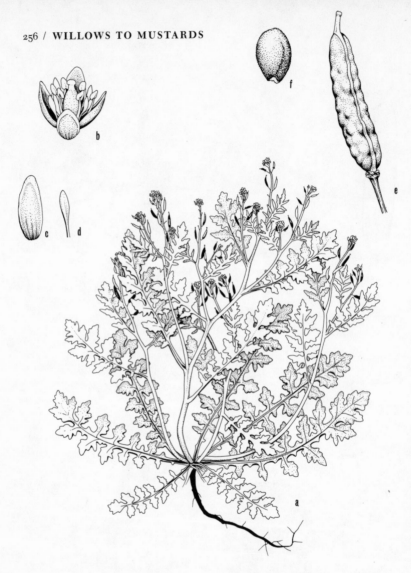

120. Rorippa truncata (Blunt-leaved Yellow Cress). *a.* Habit, X½. *b.* Flower, X7½. *c.* Sepal, X10. *d.* Petal, X10. *e.* Fruit, X10. *f.* Seed, X15.

ascending pedicels 3 mm long or longer, the seeds up to 75 per silique.

Rorippa islandica is interpreted in this work to include taxa previously known as *Rorippa palustris* (L.) Bess. and *Rorippa hispida* (Desv.) Britt.

Three varieties have been collected in Illinois.

1. Stems and leaves glabrous or nearly so _____ 2
1. Stems and leaves hirsutulous _____ 5c. *R. islandica* var. *hispida*
  2. Leaves pinnately compound or deeply pinnatifid, membranaceous
  _____ 5a. *R. islandica* var. *islandica*
  2. Leaves coarsely or shallowly toothed, firm _____
  _____5b. *R. islandica* var. *fernaldiana*

5a. Rorippa islandica (Oeder) Borbás var. **islandica** *Fig. 121.*
Sisymbrium amphibium L. var. *palustre* L. Sp. Pl. 657. 1753.
Radicula palustris (L.) Moench, Meth. 263. 1794.
Nasturtium palustre (L.) DC. Syst. 2:191. 1821.
Rorippa palustris (L.) Bess. Enum. 27. 1821.

Stems and leaves glabrous or nearly so; leaves pinnately compound or deeply pinnatifid, membranaceous.

COMMON NAME: Marsh Yellow Cress.
HABITAT: Mud and sand flats.
RANGE: Quebec to Michigan, south to Missouri and New Jersey; Greenland; Europe; Asia.
ILLINOIS DISTRIBUTION: Jackson Co.: along Mississippi River at Grand Tower, *D. K. Evans.*
This variety, which generally ranges north of Illinois, is known only in Illinois from mud and sand flats along the Mississippi River near Grand Tower.
  The distinguishing feature of this variety lies in the deep divisions of all the leaves.
  The flowers are formed from July to September.

5b. Rorippa islandica (Oeder) Borbás var. **fernaldiana** Butt. & Abbe, Rhodora 42:28. 1940. *Fig. 122a–f.*

Stems and leaves glabrous or nearly so; leaves coarsely or shallowly toothed, firm.

COMMON NAME: Marsh Yellow Cress.
HABITAT: Muddy shores.
RANGE: Labrador to British Columbia, south to California, Texas, Louisiana, and Virginia.
ILLINOIS DISTRIBUTION: Common throughout the state.
This is the most abundant of the three varieties of *R. islandica* in Illinois. It occurs along most muddy shores in the state.

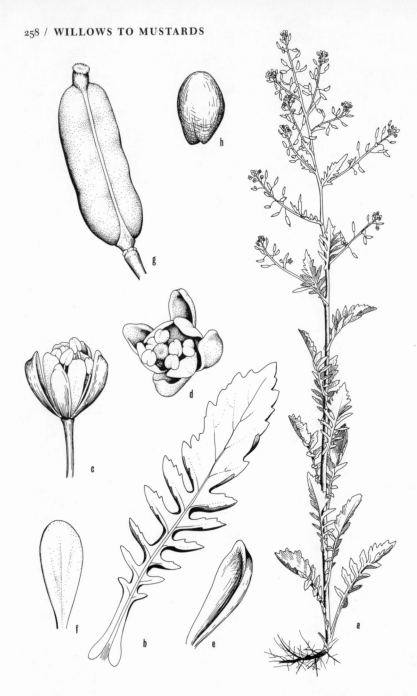

121. *Rorippa islandica* var. *islandica* (Marsh Yellow Cress).   *a.* Habit, X¼.   *b.* Leaf, X1.   *c.* Flower, X10.   *d.* Flower (top view), X10.   *e.* Sepal, X20.   *f.* Petal, X20.   *g.* Fruit, X7½.   *h.* Seed, X25.

122. *Rorippa islandica* var. *fernaldiana* (Marsh Yellow Cress). *a*. Habit, X¼. *b*. Flower, X10. *c*. Sepal, X15. *d*. Petal, X15. *e*. Fruit, X6. *f*. Seed, X15. var. *hispida*. *g*. Stem, X1½. *h*. Fruit, X6.

The glabrous herbage and the toothed leaves distinguish this variety from the others in Illinois.

The flowers appear from May to November.

5c. Rorippa islandica (Oeder) Borbás var. **hispida** (Desv.) Butt. & Abbe, Rhodora 42:26. 1940. *Fig. 122g–h.*

Brachylobus hispidus Desv. Journ. Bot. 3:183. 1814.

Nasturtium hispidum (Desv.) DC. Syst. 2:201. 1821.

Nasturtium palustre (L.) DC. var. *hispidum* (Desv.) Gray, Man. ed. 2, 30. 1856.

Rorippa hispida (Desv.) Britt. Mem. Torrey Club 5:169. 1894.

Radicula hispida (Desv.) Britt. Torreya 6:30. 1906.

Radicula palustris (L.) Moench var. *hispidum* (Desv.) B. L. Robins. Rhodora 10:32. 1908.

Stems and leaves hirsutulous.

COMMON NAME: Hairy Marsh Yellow Cress.

HABITAT: Wet ground near ponds.

RANGE: Newfoundland to British Columbia, south to New Mexico, Texas, and Florida; West Indies.

ILLINOIS DISTRIBUTION: Known only from Boone, Cook, and McHenry counties.

This rare taxon, with hirsutulous leaves and stems, is sometimes considered to be a distinct species. Since the major difference is in the presence of pubescence, the varietal status is more appropriate.

The flowers occur from May to October.

35. *Chorispora* DC.–Chorispora

Annual or perennial herbs; leaves pinnatifid or entire, basal and cauline; inflorescence racemose, the flowers purple or yellow, actinomorphic, bractless; sepals 4, free; petals 4, free; stamens 6; siliques constricted between the seeds, beaked.

This genus is composed of 10 Asian species.

Only the following adventive species occurs in Illinois.

1. **Chorispora tenella** DC. Syst. 2:435. 1821. *Fig. 123.*

Annual herb from a long slender root; stems erect to ascending, branched, sparsely glandular-hairy to nearly glabrous, to 0.5 m tall; leaves oblong to lanceolate, acute at the apex, cuneate to the petiolate base, sinuate-dentate, to 8 cm long, to 2 cm broad, usually sparsely glandular-hairy, the lowermost leaves sometimes runci-

123. *Chorispora tenella* (Chorispora).   *a*. Habit, X½.   *b*. Flower, X4.   *c*. Sepal, X9.   *d*. Petal, X7½.   *e*. Fruit, X1½.   *f*. Seed, X25.

nate; inflorescence racemose, the flowers few to several, purple, about 1.2 cm long; sepals 4, green, linear-lanceolate, 3–6 mm long; petals 4, purple, clawed, 6–12 mm long; siliques ascending, arcuate, glandular-hairy, to 4 cm long, to 3 mm broad, constricted slightly between the seeds, dehiscing transversely into 2-seeded segments, with a subulate beak up to ⅓ the length of the entire fruit, the seeds very narrowly winged or wingless.

COMMON NAME: Chorispora.

HABITAT: Disturbed areas, particularly along railroads.

RANGE: Native to Asia; adventive mostly in the western United States, but uncommon eastward.

ILLINOIS DISTRIBUTION: Scattered in much of Illinois, but very sporadic.

This brightly flowered adventive was unknown in Illinois until about 1974. Then, for a period of two or three years, it was collected several times throughout much of Illinois.

By 1978, it had disappeared from most of the original Illinois stations.

Chorispora tenella flowers from late March to August.

36. *Coronopus Rupp. ex L.* – Wart Cress

Annual or biennial, ill-smelling herbs; leaves pinnatifid, inflorescence racemose, axillary; sepals 4, free; petals 4, free, very small; stamens 2, 4, or 6; ovary superior, 2-locular; fruit a bilobed silicle, each lobe with one seed.

There are eight species in this genus, all native to the Old World.

Only the following species occurs in Illinois.

1. **Coronopus didymus** (L.) Sm. Fl. Brit. 3:691. 1800. *Fig. 124.*
Lepidium didymum L. Mant. 92. 1767.

Annual or biennial, ill-smelling herbs from long, slender roots; stems spreading, matted, somewhat pubescent, up to 40 cm long; all leaves deeply pinnatifid, sparsely pubescent, up to 3 cm long, the segments entire or toothed, the lowermost leaves petiolate, the uppermost sessile; inflorescence densely racemose, axillary, to 3 cm long; flowers minute, white, on slender pedicels up to 2 mm long; sepals 4, green, to 2 mm long; petals 4, white, 1–2 mm long; stamens often only 2; fruit a bilobed silicle notched at the apex, each lobe hemispherical, about 1.5 mm long and broad, roughly wrinkled, 1-seeded.

124. Coronopus didymus (Wart Cress). *a.* Upper part of plant, X½. *b.* Habit (in silhouette), X¼. *c.* Flower, X7½. *d.* Sepal, X10. *e.* Petal, X10. *f.* Fruit, X10. *g.* Seed, X20.

COMMON NAME: Wart Cress.
HABITAT: Waste ground.
RANGE: Native of Europe; adventive in the United States, particularly in the western and southern states.
ILLINOIS DISTRIBUTION: Known only from Cook County.
The wart cress is distinguished by its minute white flowers borne in dense racemes and by its bilobed, strongly wrinkled fruits.
This species flowers from May to September.

37. *Matthiola R. Br.* – Stock

Perennial herbs; leaves alternate, simple; inflorescence racemose, with several relatively large, showy flowers; sepals 4, green, free; petals 4, brightly colored, free; stamens 6; ovary superior, 2-locular; fruit a silique, with flattened seeds.

Matthiola, a genus composed of several species, is native to the Old World.

Only the following species, escaped from cultivation, has been found in Illinois.

1. **Matthiola incana** (Willd.) R. Br. in Ait. Hort. Kew. ed. 2, 4:119. 1812. *Fig. 125.*
Cheiranthus incanus Willd. Sp. Pl. 3:520. 1800–1803.

Perennial herbs from elongated roots; stems erect, to 0.5 m tall, much branched, pubescent; inflorescence racemose, densely flowered; flowers up to 2 cm broad, purple, varying to pink or even white; sepals 4, green, to 1 cm long; petals 4, to 2 cm long; siliques linear, flattened, to 12 cm long, with several flattened, winged seeds.

COMMON NAME: Stock.
HABITAT: Disturbed ground.
RANGE: Native of Europe; rarely escaped from gardens in the United States.
ILLINOIS DISTRIBUTION: Known only from Grundy and Kendall counties.
Stock is a favorite in perennial flower gardens and is a rare escape from cultivation.

125. *Matthiola incana* (Stock). *a.* Habit, X¼. *b.* Flower, X2½. *c.* Sepal, X3¾. *d.* Petal, X2¼. *e.* Fruit, X1½. *f.* Seed, X25.

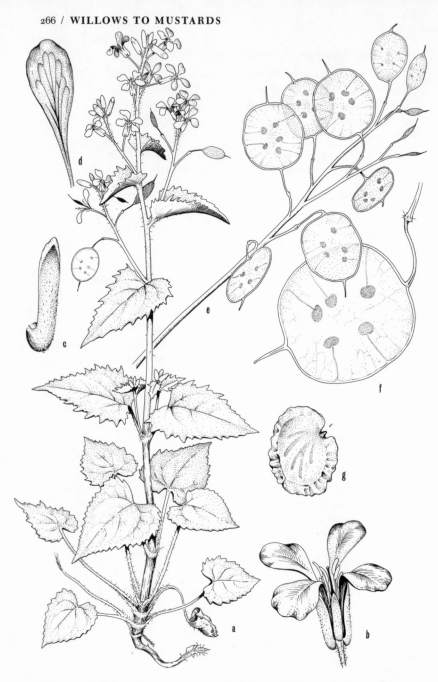

126. *Lunaria annua* (Money-plant). *a.* Habit, X¼. *b.* Flower, X2½. *c.* Sepal,
X5. *d.* Petal, X2½. *e.* Fruiting raceme, X½. *f.* Fruit, X1. *g.* Seed, X5.

38. *Lunaria* [*Tourn.*] L. —Honesty

Annual or perennial, pubescent herbs; leaves alternate, simple, toothed; inflorescence of terminal racemes; flowers several, large, purple; sepals 4, free; petals 4, free, clawed; stamens 6; ovary superior; siliques flattened, reticulate, long-stipitate, with a silvery replum, with a few large, winged seeds.

Three European species comprise this genus.

Only the following escape from cultivation occurs in Illinois.

1. Lunaria annua L. Sp. Pl. 653. 1753. *Fig. 126.*

Annual or biennial herbs from a stout root; stems erect, pubescent, sparingly branched, to 1 meter tall; leaves broadly ovate to ovate-lanceolate, pubescent, coarsely dentate, the lowermost up to 15 cm long and long-petiolate, the uppermost progressively smaller and sessile; inflorescence terminal, racemose, several-flowered; flowers up to 2 cm across, purple, on slender, ascending pedicels; sepals 4, green, lanceolate, with two of them slightly swollen at the base, to 6 mm long; petals 4, clawed, to 15 mm long; siliques oval to nearly orbicular, flat, to 5 cm long, to 3 cm broad, the valves papery, reticulate, falling away to expose the persistent, silvery replum; seeds few, reniform.

COMMON NAME: Money-plant; Silver Dollar Plant; Honesty.

HABITAT: Disturbed areas.

RANGE: Native of Europe; rarely escaped from gardens in the United States.

ILLINOIS DISTRIBUTION: Known from Champaign County. This coarse, pretty-flowered mustard is frequently cultivated for its unique fruits which, after the valves fall away, show the large, transparent, silvery replum.

Money-plant flowers during the second year, during May and June.

Species and Varieties Excluded

Arabis confinis S. Wats. This binomial has been attributed by several authors to Illinois plants which are actually *A. drummondii*.

Barbarea stricta Andrz. Pepoon first reported this species in 1927, but apparently this was an error for *B. vulgaris*.

Brassica napobrassica Mill. The rutabaga is sometimes cultivated, but I have not seen any specimens marked as escapes from cultivation.

Cardamine rotundifolia Michx. Mead (1846) reported this species from Illinois, although the material he was familiar with was probably *C. douglassii*.

Cleome speciosissima Deppe. The McCree specimen from Jackson County, first reported by Jones et al. as *C. speciosissima*, has been annotated by Iltis as *C. hassleriana*. The true *C. speciosissima* apparently does not occur in Illinois.

Cleome spinosa L. Reports of this cultivated species by many writers from Illinois are really of *C. hassleriana*, according to the annotations of Iltis.

Dentaria diphylla Michx. Huett (1897) undoubtedly should have been reporting *D. laciniata* from Illinois, rather than this southeastern species.

Descurainia pinnata (Walt.) Britt. Kibbe's (1952) report of this species from Hancock County is an error for *D. brachycarpa*.

Gynandropsis gynandra (L.) Briq. This species, reported in 1857 by Lapham as *Synandropsis pentaphylla*, has not been verified during this study. It was reputedly a garden escape.

Lepidium apetalum Willd. Gates (1912) and Pepoon (1927) reported this species from northeastern Illinois, but the specimens they had must surely have been *L. densiflorum*.

Lesquerella nuttallii (Torr. & Gray) S. Wats. This species does not occur in Illinois, although Pepoon (1927) reported it from Cook County in error for *L. gracilis*.

Lesquerella spathulata Rydb. When Gleason first discovered *L. ludoviciana* in 1904, he thought it was the western *L. spathulata*.

Populus angulata Ait. Mead's (1846), Lapham's (1857), and Patterson's (1876) references to this species should actually have been for *P. deltoides*.

Populus × jackii Sarg. Specimens called this from Cook County seem to be nothing more than *P. balsamifera*. *Populus × jackii* is a reputed hybrid between *P. balsamifera* and *P. deltoides*.

Rorippa obtusa (Nutt.) Britt. Stuckey (1966) has indicated that the Illinois specimen originally identified as this is actually *R. truncata*.

Salix coactilis Fern. Pepoon's recording of this plant from Illinois in 1917 and again in 1927 is based on a misidentification for *S. sericea*.

Salix myrtilloides L. This Linnaean species was first attributed to Illinois by Bebb in 1876, and later by Brendel in 1887. These reports were in error for *Salix pedicellaris*.

Sisymbrium canescens Nutt. Until the binomial *Descurainia brachycarpa* was proposed by O. E. Schulz, the tansy mustard was known as *S. canescens*, which is an entirely different species.

Sisymbrium incisum Engelm. var. *filipes* A. Gray. This variety was attributed to Hancock County by Kibbe (1952), but the report is an error for *Descurainia brachycarpa*.

Sisymbrium irio L. Pepoon (1927) reported that Bebb collected this mustard along a railroad in Cook County, but I have seen no specimens to substantiate this report.

Sophia incisa (Engelm.) Greene. This species does not occur in Illinois, although Gates (1926) applied this binomial to a specimen of *Descurainia brachycarpa*.

Summary of the Taxa Treated in This Volume

Families	Genera	Species	Lesser Taxa	Hybrids
Salicaceae	2	31	5	4
Tamaricaceae	1	1		
Capparidaceae	2	4	1	
Resedaceae	1	1		
Brassicaceae	38	80	11	
Totals	44	117	17	4

GLOSSARY
LITERATURE CITED
INDEX OF PLANT NAMES

GLOSSARY

Actinomorphic. Having radial symmetry; regular, in reference to a flower.

Acuminate. Gradually tapering to a point.

Acute. Sharply tapering to a point.

Alliaceous. Having the odor of an onion.

Ament. A spike of unisexual, apetalous flowers; a catkin.

Annual. Living only for one year.

Apetalous. Without petals.

Appressed. Lying flat against the surface.

Arcuate. Curved.

Areolate. Possessing a network of closed veins.

Attenuate. Gradually becoming narrowed.

Auriculate. Bearing an earlike process.

Biennial. Living only for two years and blooming the second year.

Bifid. Two-cleft.

Bifurcate. Forked.

Bipinnate. Divided once into distinct segments, with each segment in turn divided into distinct segments.

Bipinnatifid. Divided part way to the center, with each lobe again divided part way to the center.

Bract. An accessory structure at the base of many flowers, usually appearing leaflike.

Bracteate. Bearing one or more bracts.

Caducous. Falling away very early.

Canescent. Grayish-hairy.

Capillary. Threadlike.

Capsule. A dry, dehiscent fruit composed of more than one carpel.

Catkin. A spike of unisexual, apetalous flowers; an ament.

Caudex (pl., **caudices**). The woody base of a perennial plant.

Cauline. Pertaining to a stem.

Ciliate. Bearing cilia.

Ciliolate. Bearing small cilia.

Clavate. Club-shaped.

Claw. A narrow, basal stalk, particularly of a petal.

Coma. A tuft of hairs at the end of a seed.

Comose. Bearing a coma, or tuft of hairs.

Compressed. Flattened.

Conduplicate. Folded together lengthwise.

Connate. Union of like parts.

Cordate. Heart-shaped.

Coriaceous. Leathery.

Corymb. A type of inflorescence where the pedicellate flowers are arranged along an elongated axis but with the flowers all attaining about the same height.

Crenate. With round teeth.

Cuneate. Wedge-shaped; tapering to the base.

Cupular. Shaped like a small cup.

Cyathiform. Cup-shaped.

Decumbent. Lying flat, but with the tip ascending.

Decurrent. Adnate to the petiole or stem and then extending beyond the point of attachment.

Dehiscent. Splitting at maturity.

Dentate. With sharp teeth, the tips of which project outward.

Denticulate. With small, sharp teeth, the tips of which project outward.

Dioecious. With staminate flowers on one plant, pistillate flowers on another.

Disk. An enlarged outgrowth of the receptacle.

Eglandular. Without glands.

Ellipsoid. Referring to a solid object which is broadest at the middle, gradually tapering to both ends.

Elliptic. Broadest at middle, gradually tapering to both ends.

Emarginate. Having a shallow notch at the extremity.

Erose. With an irregularly notched margin.

Falcate. Sickle-shaped.

Fibrous. Referring to roots borne in tufts.

Filament. That part of a stamen supporting the anther.

Filiform. Threadlike.

Fimbriate. Fringed.

Fusiform. Spindle-shaped; tapering at both ends.

Gibbous. Swollen on one side.

Glabrate. Becoming smooth.

Glabrous. Without pubescence or hairs.

Glaucous. With a whitish covering which can be rubbed off.

Globose. Round; globular.

Granular. Having a grainy appearance.

Gynophore. The stipe of a pistil.

Hirsute. With stiff hairs.

Hirsutulous. With minute stiff hairs.

Hirtellous. Finely hirsute.

Hispid. With rigid hairs.

Hispidulous. With short, rigid hairs.

Hoary. Grayish-white, usually referring to pubescence.

Hyaline. Transparent.

Indehiscent. Not splitting open at maturity.

Inflorescence. A cluster of flowers.

Keeled. Bearing a ridgelike process.

Lacerate. Torn, forming a very jagged margin.

Laciniate. Divided into narrow, pointed divisions.

Lanceolate. Lance-shaped; broadest near the base, gradually tapering to the narrower apex.

Lanceoloid. Referring to a solid object which is broadest near the base, gradually tapering to the narrower apex.

Lenticel. Corky openings on bark of twigs and branches.

Linear. Elongated and uniform in width throughout.

Locule. The cavity of an ovary.

Lyrate. Pinnatifid, with the terminal lobe much larger than the lower ones.

Marginate. Possessing a margin, often in the form of a narrow wing.

Mucronulate. Possessing a very short, abrupt tip.

Obcordate. Inversely heart-shaped.

Oblanceolate. Reverse lance-shaped; broadest at apex, gradually tapering to narrow base.

Oblong. Broadest at the middle, and tapering to both ends, but broader than elliptic.

Oblongoid. Referring to a solid object which, in side view, is nearly the same width throughout.

Obovate. Broadly rounded at the apex, becoming narrowed below.

Obovoid. Referring to a solid object which is broadly rounded at the apex, becoming narrowed below.

Obtuse. Rounded at the apex.

Orbicular. Round.

Oval. Broadly elliptic.

Ovate. Broadly rounded at base, becoming narrowed above; broader than lanceolate.

Ovoid. Referring to a solid object which is broadly rounded at the base, becoming narrowed above.

Ovule. A structure within the ovary which produces the egg and, following fertilization, develops into the seed.

Panicle. A type of inflorescence composed of several racemes.

Paniculate. Bearing a panicle.

Papillose. Bearing pimplelike processes.

Parietal. Borne on the wall.

Pedicel. The stalk of a flower.

Pendulous. Hanging.

Perennial. Living more than two years.

Perfoliate. Referring to a leaf which appears to have the stem pass through it.

Perianth. Those parts of a flower including both the sepals and petals.

Petiolate. Bearing a petiole, or leafstalk.

Petiole. The stalk of a leaf.

Petiolulate. Bearing a petiolule, or leaflet stalk.

Pilose. Bearing soft hairs.

Pinnate. Divided once into distinct segments.

Pinnatifid. Said of a simple leaf or leaf-part which is cleft or lobed only part way to its axis.

Pistillate. Bearing pistils but not stamens.

Placenta. The point of attachment of the ovary to the ovary wall.

Placentation. Referring to the way the ovule is attached to the ovary wall.

Polypetaly. With free petals.

Puberulent. With minute hairs.

Pubescent. Bearing some kind of hairs.

Raceme. A type of inflorescence where pedicellate flowers are arranged along an elongated axis.

Rachis. A primary axis.

Reniform. Kidney-shaped.

Repand. Wavy along the margin.

Replum. A membranous partition between the two valves of the fruit of the Brassicaceae.

Resinous. Producing a sticky secretion, or resin.

Reticulate. Resembling a network.

Retrorse. Pointing downward.

Revolute. Rolled under from the margin.

Rhizome. An underground, horizontal stem, bearing nodes, buds, and roots.

Rhombic. Becoming quadrangular.

Rosette. A cluster of leaves in

a circular arrangement at the base of a plant.

Rugose. Wrinkled.

Runcinate. Sharply cut, with the divisions pointing backward.

Sagittate. Shaped like an arrowhead.

Scabrous. Rough to the touch.

Scape. A leafless stalk bearing a flower or inflorescence.

Scarious. Thin and membranous.

Septum. A dividing wall.

Sericeous. Silky; bearing soft, appressed hairs.

Serrate. With teeth which project forward.

Serrulate. With very small teeth which project forward.

Sessile. Without a stalk.

Setose. Bearing bristles.

Silicle. A short silique.

Silique. An elongated capsule with a central partition separating the valves.

Sinuate. Wavy along the margins.

Spatulate. Oblong, but with the basal end elongated.

Staminate. Bearing stamens but not pistils.

Stellate. Star-shaped.

Stipe. A stalk.

Stipitate. Possessing a stalk.

Stolon. A slender, horizontal stem on the surface of the ground.

Stoloniferous. Bearing stolons.

Strigose. With appressed, straight hairs.

Subcordate. Nearly heart-shaped.

Subcuneate. Nearly wedge-shaped.

Subentire. Nearly entire.

Suborbicular. Nearly spherical.

Subulate. With a very short, narrow point.

Superior. Referring to the position of the ovary when the free floral parts arise below the ovary.

Terete. Round, in cross section.

Tomentose. Pubescent with matted wool.

Tomentulose. Finely pubescent with matted wool.

Torulose. With small constrictions.

Trifoliolate. Bearing three leaflets.

Tripinnate. Divided three times into distinct segments.

Truncate. Abruptly cut across.

Tuber. An underground fleshy stem formed as a storage organ at the end of a rhizome.

Turgid. Swollen to the point of bursting.

Undulate. Wavy.

Unifoliolate. Having one leaflet.

Unilocular. Having a single locule.

Valvate. Opening by the splitting of the valves.

Valve. That part of a capsule which splits.

Villous. With long, soft, slender, unmatted hairs.

Virgate. Wandlike.

Viscid. Sticky.

Whorl. An arrangement of three or more structures at any given point.

Zygomorphic. Bilaterally symmetrical.

LITERATURE CITED

Argus, G. U. 1964. Preliminary notes on the flora of Wisconsin. No. 51. Salicaceae. The genus *Salix*—the willows. Transactions of the Wisconsin Academy of Sciences, Arts and Letters 53:217–72.

Ball, C. R. 1932. Salicaceae. In C. C. Deam, Shrubs of Indiana, 2d ed. Indianapolis: Indiana Department of Conservation, pp. 37–70.

———. 1948. *Salix petiolaris* J. E. Smith: American, not British. Bulletin of the Torrey Botanical Club 75:178–87.

Bebb, M. S. 1876. Salix. In Patterson, Catalogue of the phaenogamous and vascular cryptogamous plants of Illinois. Privately printed for the author at Oquawka, Ill. 54 pp.

Brendel, F. 1887. Flora Peoriana. Privately printed for the author at Peoria, Ill. 89 pp.

Brown, W. H. 1938. The bearing of nectaries on the phylogeny of flowering plants. Proceedings of the American Philosophical Society 79:549–95.

Costello, D. F. 1935. Preliminary reports on the flora of Wisconsin. No. 35. Salicaceae. Transactions of the Wisconsin Academy of Sciences, Arts and Letters 29:299–318.

Cronquist, A. 1968. The evolution and classification of flowering plants. Boston: Houghton Mifflin Co. 396 pp.

Crosswhite, F. S., and H. H. Iltis. 1966. Studies in the Capparidaceae X. Orthography and conservation: Capparidaceae vs. Capparaceae. Taxon 15(6):205–14.

Ernst, W. R. 1963. The genera of Capparaceae and Moringaceae in the southeastern United States. Journal of the Arnold Arboretum 44:81–95.

Feldman, A. W. 1942. Trees and shrubs of Champaign County, Illinois. Transactions of the Illinois Academy of Science 35:60–61.

Fernald, M. L. 1922. Some variations of *Cakile edentula*. Rhodora 24:21–23.

———. 1946. Technical studies on North American plants. II. Difficulties in North American *Salix*. Rhodora 48:27–40, 41–49.

———. 1950. Gray's manual of botany. 8th ed. New York: American Book Co. 1632 pp.

Gates, F. C. 1912. The vegetation of the beach area in northeastern Illinois and southeastern Wisconsin. Bulletin of the Illinois State Laboratory of Natural History 9:255–372.

———. 1926. Contributions to the flora of Hancock County, Illinois. Transactions of the Illinois Academy of Science 182:225–34.

Gleason, H. A. 1910. The vegetation of the inland sand deposits of Illinois. Bulletin of the Illinois State Laboratory of Natural History 9:23–174.

———. 1952. The new Britton and Brown illustrated flora of the northeastern United States and adjacent Canada. Vol. 2. New York: New York Botanical Garden. 655 pp.

Higley, W. K., and C. S. Raddin. 1891. Flora of Cook County, Illinois, and a part of Lake County, Indiana. Bulletin of the Chicago Academy of Science 2:1–168.

Hopkins, M. 1937. *Arabis* in eastern and central North America. Rhodora 39:63–98, 106–48, 155–86.

Huett, J. W. 1897. Essay toward a natural history of La Salle County, Illinois. Flora La Sallensis. Part 1. Privately printed for the author at Ottawa, Ill. 136 pp.

Iltis, H. H. 1954. Studies in the Capparidaceae. I. *Polanisia dodecandra* (L). DC., the correct name for *Polanisia graveolens* Rafinesque. Rhodora 56:65–70.

———. 1958. Studies in the Capparidaceae. IV. *Polanisia* Raf. Brittonia 10:33–58.

———. 1966. Studies in the Capparidaceae. VIII. *Polanisia dodecandra* (L.) DC. Rhodora 68:41–47.

Jones, G. N. 1963. Flora of Illinois. 3d ed. South Bend: University of Notre Dame Press. 402 pp.

———; G. D. Fuller; G. S. Winterringer; H. E. Ahles; and A. Flynn. 1955. Vascular plants of Illinois. Urbana: University of Illinois Press, and the Illinois State Museum, Springfield. 593 pp.

Kibbe, A. L. 1952. A botanical study and survey of a typical midwestern county (Hancock County, Illinois). Privately published by the author at Carthage, Illinois. 425 pp.

Lapham, I. A. 1857. Catalogue of the plants of the state of Illinois. Transactions of the Illinois State Agricultural Society 2:429–550.

Lindsey, A.; P. Petty; D. Sterling; and W. VanAsdall. 1961. Vegetation and environment along the Wabash and Tippecanoe rivers. Ecological Monographs 31:105–56.

McVaugh, R. 1949. New and adopted names. 3. Questionable validity of names published in Gilibert's flora of Lithuania. Gentes Herbarium 8:83–90.

Mead, S. B. 1846. Catalogue of plants growing spontaneously in the state of Illinois, the principal part near Augusta, Hancock County. Prairie Farmer 6:35–36, 60, 93, 119–22.

Mohlenbrock, R. H. 1975. Guide to the vascular flora of Illinois. Carbondale: Southern Illinois University Press. 494 pp.

Patman, J. P., and H. H. Iltis. 1961. Preliminary reports on the flora of Wisconsin. No. 44. Cruciferae—mustard family. Transactions of the Wisconsin Academy of Sciences, Arts and Letters 50:17–72.

Patterson, H. N. 1876. Catalogue of the phaenogamous and vascular cryptogamous plants of Illinois. Privately printed for the author at Oquawka, Ill. 54 pp.

Pepoon, H. S. 1927. An annotated flora of the Chicago area. Bulletin of the Chicago Academy of Science 8:1–554.

Raup, H. M. 1943. The willows of the Hudson Bay region and the Labrador Peninsula. Sargentia 4:81–135.

Rollins, R. C. and E. A. Shaw. 1973. The genus *Lesquerella* (Cruciferae) in North America. Cambridge: Harvard University Press. 282 pp.

Rossbach, G. B. 1958. New taxa and new combinations in the genus Erysimum in North America. Aliso 4:115–24.

Schneider, C. K. 1921. Notes on American willows. XI. Journal of the Arnold Arboretum 1:185–204.

Steyermark, J. A. 1963. Flora of Missouri. Ames: Iowa State University Press. 1725 pp.

Stuckey, R. L. 1966. *Rorippa walteri* and *R. obtusa* synonyms of *R. teres* (Cruciferae). Sida 2:409–18.

Swink, F. 1974. Plants of the Chicago region. 2d ed. Lisle, Ill.: Morton Arboretum. 474 pp.

Tehon, L. R. 1942. Fieldbook of native Illinois shrubs. Urbana: Illinois Natural History Survey, Manual No. 3. 307 pp.

Thorne, R. F. 1968. Synopsis of a putatively phylogenetic classification of flowering plants. Aliso 6:57–66.

Wagner, W. H. 1970. The Barnes Hybrid Aspen, *Populus* × *barnesii* hybr. nov., a nomenclatural case in point. Michigan Botanist 9:53–54.

Wheeler, L. C. 1938. The names of three species of *Brassica*. Rhodora 40:306–9.

Youngberg, A. D. 1970. *Salix starkeana* in North America. Rhodora 72:548–50.

INDEX OF PLANT NAMES